THE YEAR OF THE GORILLA

THE YEAR OF THE
Gorilla

GEORGE B. SCHALLER

WITH LINE DRAWINGS BY THE AUTHOR

THE UNIVERSITY OF CHICAGO PRESS

ISBN: 0–226–73637–7 (clothbound); 0–226–73638–5 (paperbound)

Library of Congress Catalog Card Number: 64–13946

THE UNIVERSITY OF CHICAGO PRESS, CHICAGO 60637

To KAY

Contents

Illustrations

MAPS

I think I could turn and live with animals, they are so placid and
    self contain'd,
I stand and look at them long and long.
They do not sweat and whine about their condition,
They do not lie awake in the dark and weep for their sins.
They do not make me sick discussing their duty to God,
Not one is dissatisfied, not one is demented with the mania of
    owning things,
Not one kneels to another, nor to his kind that lived thousands
    of years ago,
Not one is respectable or unhappy over the whole earth.

"Song of Myself"
WALT WHITMAN

INTRODUCTION

Chance remarks and luck have often played an important role in my life, and the way in which I came to study gorillas was no exception. In January, 1957, I entered the office of Dr. John T. Emlen, professor of zoölogy at the University of Wisconsin, to ask a question. At that time I was one of his graduate students working in bird behavior.

Doc leaned back in his chair and somewhat jokingly asked: "Would you like to study gorillas?"

"Sure," I replied impulsively.

Dr. Emlen told me that Harold Coolidge, of the National Academy of Sciences, was eager to launch an expedition to study the behavior of free-living gorillas. Coolidge is, I later found out, one of those exceptional scientists who are also good organizers. The idea of studying gorillas had occurred to him in 1927, when, as a member of the Harvard African Expedition, he had visited gorilla country in Central Africa. Unable to study the animals himself, he was still looking for someone to do it. I went home and the same evening wrote him a letter.

The path from an office in Wisconsin to the forests of Africa is not a straight one. Many letters have to be written, a sponsor has to be found, the help of foreign institutions has to be sought, and, most importantly, money has to be obtained. One of my first tasks was to learn what had been written about the creature I proposed to study.

Probably no animal has fired the imagination of man to the same extent as has the gorilla. Its manlike appearance and tremendous strength, its remote habitat and reputed belligerence, have endowed the beast with a peculiar fascination and stirred popular and scientific interest. It appears to possess some transcendent quality which inspires every visitor to its realm to put his experiences in print. I read through literally hundreds of popular books, articles, and newspaper stories, and I examined scientific papers and glanced through textbooks. If mere verbiage were a measure of knowledge there would be little left to study. Unfortunately, the serious researcher must discard most published information about the behavior of free-living gorillas. Much of it is sensational, irresponsible and exaggerated prevarications, with little concern for truth. The gorilla is usually pictured as a ferocious and bloodthirsty beast with an amazing array of human and superhuman traits, all basically treacherous. Another common piece of literature is the intrepid hunter's tale, which may contain useful information, but, since the ape is studied along the sights of a rifle, the value of the tale is curtailed. Such accounts usually include a photograph of a gorilla, shot through the head and propped against a tree with our hero squatting next to the huge cadaver. Then, of course, there is the innocuous adventurer, who, camera in hand, crashes through gorilla land with a long string of porters at his heels. The finding of a gorilla nest or perhaps the brief sighting of the ape itself makes him an expert on all aspects of the gorilla's life history. He writes a learned article or even a book, but to compensate for his lack of knowledge he includes native tales, rumors, and statements from older literature, no matter how dubious. Much of such questionable information about gorillas has been copied and recopied so often that through mere repetition it has achieved the ring of truth.

As I first began to read, I did not know what was true and what was false, but early in my searching I became aware that, if I discarded all generalizations not based on supporting facts and all subjective interpretations, I had hardly any concrete information to go on. There were naturally some conspicuous exceptions, and from these I learned about the history of the discovery of the gorilla, its taxonomy, and the broad outlines of the ape's behavior. Yet, I thought, here is a creature, considered with the chimpanzee the nearest relative of man, and we know almost nothing about its life in the wild. Does it live in small family units or in large groups; how many males and females are there in each group; what do groups do when they meet; how far do they travel each day; how long are infants dependent on their mothers? These and many other basic questions were unknown in 1957. In an age when man more than ever before is wondering about his origins and worrying about his behavior, in an age when he is reaching for the moon, he is just beginning to study his closest relations and himself. As the German dramatist Friedrich Hebbel stated: "It would be good if man concerned himself more with the history of his nature than that of his deeds."

The eastern or lowland gorilla (*Gorilla gorilla gorilla*) inhabits West Africa from southern Nigeria southward through Cameroun, Gabon, Rio Muni, and almost to the Republic of the Congo. From the coast, its range extends inland about five hundred miles to the vicinity of the Ubangi River, a tributary of the Congo River, a vast area of flatlands and hills densely covered with rain forest. Few Europeans penetrated these hot and humid jungles until the present century, with the result that the gorilla, the largest of the apes, was the last ape to become known to science.

In 470 B.C., Hanno set out from Carthage with a grand expedition consisting of sixty fifty-oared galleys loaded with colonists and merchandise. In the foothills of the mountains of Sierra Leone, the colonists met some hairy sylvan creatures who, when attacked, threw stones at them. Three of the animals, called "gorillai," were captured. Pliny relates that at the time of the Roman invasion in 146 B.C., two of the skins were still preserved at Carthage in the temple of Astarte. Although this is the first historical use of the term "gorilla," the animals in question were probably baboons or chimpanzees.

In 1559, Andrew Battell, an English adventurer, was taken prisoner by the Portuguese. He was enrolled in their colonial troops in West Africa and spent several years near the Mayombe River. There he met two kinds of apes, and their description was published in 1625 in an obscure book entitled *Purchas his Pilgrimes*. One of the apes, the *pongo*, is undoubtedly the gorilla.

This Pongo is in all proportions like a man; but that he is more like a giant in stature, than a man: for he is very tall, and hath a man's face, hollow-eyed, with long haire upon his browes. His face and eares are without haire, and his hands also. His bodie is full of haire, but not very thicke, and it is of a dunnish colour. He differeth not from a man, but in his legs; for they have no calfe. . . . They sleepe in the trees, and build shelters for the raine.

Although Battell was the first European to hear of the gorilla, his narrative was ignored. In 1774, Lord Monboddo received a letter from a sea captain which described an ape, probably the gorilla: "This wonderful and frightful production of nature walks upright like a man; is from 7 to 9 feet high, when at maturity, thick in proportion and amazingly strong. . . ." Thomas Bowdich published in 1819 a book called *Mission from Cape Coast Castle to Ashanti* in which he discusses several apes from the Gabon, among them the *ingenu*, which is generally five feet tall and four feet broad. And three years later George Maxwell mentions a great ape surpassing the chimpanzee in size. But the credit for the actual discovery of the gorilla belongs to two missionaries, Wilson and Savage by name. When Savage was visiting Wilson at the Gabon River in 1846 he saw in the house "a skull represented by the natives to be a monkey-like animal, remarkable for its size, ferocity, and habits." Both men collected several skulls in the ensuing months and these they sent to the eminent anatomists Jeffries Wyman and Richard Owen. Savage included a description of the life and habits of the gorilla, an account which set the descriptive pattern for the next hundred years.

They are exceedingly ferocious, and always offensive in their habits, never running from man as does the Chimpanzee. . . . It is said that when the male is first seen he gives a terrific yell that resounds far and wide through the forest, something like *kh-ah!* prolonged and shrill. . . . The females and young at the first cry quickly disappear; he then approaches the enemy in great fury, pouring out his cries in quick succession. The hunter awaits his approach with gun extended; if his aim is not sure he permits the animal to grasp the barrel, and as he carries it to his mouth he fires; should the gun fail to go off, the barrel is crushed between his teeth, and the encounter soon proves fatal to the hunter.

Not to be outdone by this tale, the anatomist Owen, briefly abandoning science for mythology, wrote in 1859:

Negroes when stealing through shades of the tropical forest become sometimes aware of the proximity of one of these frightfully formidable apes by the sudden disappearance of one of their companions, who is hoisted up into the tree, uttering, perhaps, a short choking cry. In a few minutes he falls to the ground a strangled corpse.

In 1856 the American explorer Paul du Chaillu arrived in West Africa. He was the first white man actually to shoot a gorilla, and his colorful and highly exaggerated account, published in 1861 in his famous *Explorations and Adventures in Equatorial Africa*, attributed even greater ferocity to the gorilla than had his predecessors. Here he describes the climax of a gorilla hunt:

And now truly he reminded me of nothing but some hellish dream creature — a being of that hideous order, half-man half-beast, which we find pictured by old artists in some representations of the infernal regions. He advanced a few steps — then stopped to utter that hideous roar again — advanced again, and finally stopped when at a distance of about six yards from us. And here, just as he began another of his roars, beating his chest in rage, we fired, and killed him.

Du Chaillu popularized the gorilla with his accounts, but he was castigated by scientists because his descriptions were regarded as fantasy. This was unfortunate, for he was basically a competent and reliable observer. Rumor has it that the publisher returned the first draft of his book because it was not lively enough. In spite of the exaggerated descriptions, Du Chaillu's account of gorillas remained as one of the most accurate for a hundred years.

Today the habits of the lowland gorillas are still largely unknown. Hunters shoot them, zoo collectors catch them, and explorers take random notes on them in passing, but only one

scientist has made a definite long-term attempt to study the ape. In 1896, Garner published a book on his experiences in West Africa. Apparently intimidated by the supposed belligerent nature of the gorilla, he built himself an iron cage in the forest, and thus protected sat day after day, waiting for gorillas to come to him. As might be expected, his success was as restricted as he. Fred Merfield, a guide, published in 1956 the best account of gorilla hunting since Du Chaillu's. A few scientists, too, have written short papers, but there is much we do not know. The lowland gorilla still awaits study.

At the turn of the century the focus on the gorilla shifted from West Africa a thousand miles to the east, to the mountainous regions of the eastern Congo and western Uganda. In this area the Albertine rift stretches like a gigantic ditch, about thirty miles wide, from the upper White Nile to the southern end of Lake Tanganyika, representing, as it were, the backbone of Africa. A series of large lakes line the bottom of the rift valley, and these include, from north to south, Lakes Albert, Edward, Kivu, and Tanganyika. Two isolated mountain massifs project from the valley floor. Between Lakes Albert and Edward rise the fabled Mountains of the Moon or the Ruwenzories to an altitude of 16,730 feet. Just to the north of Lake Kivu, the chain of eight Virunga Volcanoes forms a huge dam across the floor of the rift valley. Both the Ruwenzori Mountains and the Virunga Volcanoes are at least partially included in the eight-thousand-square-mile Albert National Park. A chaotic jumble of mountains, reaching a height of ten thousand feet, borders the rift valley for much of its length.

It was in this spectacular and largely unknown area that rumors of the presence of a large ape reached the early explorers of Central Africa. In November, 1861, Speke and Grant traveled northward near the eastern border of what is now Rwanda and Burundi, the first Europeans to penetrate this region in search of the source of the Nile. They were told of man-like monsters, "who could not converse with men," living in the mountains toward the west. In 1866, Livingstone walked from Ujiji, an Arab slave trading post on the shores of Lake Tanganyika, westward to Nyangwe, a town which the Arabs had established on the upper Congo River in about 1860. Along the way he observed natives as they battled with creatures which he called gorillas, but which were undoubtedly chimpanzees. The explorer Stanley expressed his belief in 1890 that gorillas existed in the northeastern Congo. E. Grogan walked from Capetown to Cairo in 1898, the first man to span the continent in this manner. While hunting elephants in the Virunga Volcanoes he "came on the skeleton of a gigantic ape." In 1902, Captain Oscar von Beringe, a German officer, traveled from Usumbura, a town on the northern end of Lake Tanganyika, northward through Ruanda-Urundi, at that time a German colony. The main purpose of the trip was to impress the native chiefs and the Belgian border posts with the military might of the German empire. On October 17, 1902, Von Beringe and a Dr. England attempted to climb to the summit of Mt. Sabinio, one of the peaks in the Virunga Volcanoes. Just after making camp on a narrow ridge at an altitude of 9,300 feet, they spotted several apes above them:

We spotted from our camp a group of black, large apes which attempted to climb to the highest peak of the volcano. Of these apes we managed to shoot two, which fell with much noise into a canyon opening to the northeast of us. After 5 hours of hard work we managed to haul up one of these animals with ropes. It was a large man-like ape, a male, about 1½ m. high and weighing over 200 pounds. The chest without hair, the hands and feet of huge size. I could unfortunately not determine the genus of the ape. He was of a previously unknown size for a chimpanzee, and the presence of gorillas in the Lake region has as yet not been determined.

This paragraph lies buried in a thoroughly uninspired account of Von Beringe's trip in *Deutsches Kolonialblatt*, a colonial newspaper. Von Beringe sent the skeleton of one of the apes to the German anatomist Matschie, who was known for "creating" ape species and subspecies with wild abandon, often on the basis of one skull alone. True to form, he noted that the gorilla collected by Von Beringe was different from the West African ones. Known today as the eastern or mountain gorilla, the subspecies has been named *Gorilla gorilla beringei* in honor of its discoverer.

The mountain gorilla resembles the lowland gorilla so much that, if only one animal is at hand, even an anthropologist would have difficulty in deciding to which race it belongs. The anatomist A. Schultz has listed thirty-four morphological differences between the two, most of them minor. For example, the mountain gorilla has longer hair and a longer palate than the lowland gorilla.

Early in our research we decided that the mountain gorilla would be more suitable for a study than the lowland one. It frequented mountains which had a temperate climate, it was supposedly on the verge of extinction, making it important to gather as much information as possible before it completely died out, and several possible study localities were known, although the distribution of the mountain gorilla as a whole remained obscure. We, therefore, placed particular emphasis on reading the literature concerning the mountain gorilla.

First on the spot after the discovery of a new animal is usually an army of museum collectors, intent on shooting the biggest and best specimen. So it was with the mountain gorilla. Between 1902 and 1925, over fifty-four gorillas were taken from the Virunga Volcanoes alone, even though this small one-hundred-and-fifty square-mile area was supposedly the last stronghold of the ape. In 1921, for example, an expedition led by Prince Wilhelm of Sweden slaughtered fourteen gorillas; an American named Burbridge shot and captured nine of them between 1922 and 1925. Carl Akeley, the famous naturalist and sculptor, shot five gorillas in 1921 for the American Museum of Natural History. Fortunately, Akeley was so impressed with his quarry and the mountains in which it lived that he urged the Belgian government to set aside a permanent sanctuary for the animals where they could live in peace and be studied by scientists. Albert National Park, established on April 21, 1925, and enlarged on July 9, 1929, to include the whole chain of the Virunga Volcanoes, is a fitting tribute not only to Akeley but also to King Albert, the only Belgian king who had the quality of greatness.

Akeley returned to the volcanoes in 1926 to study the mountain gorilla, not to shoot it. He died at the beginning of the expedition and was buried in the sanctuary he helped create. H. Bingham, a laboratory psychologist, arrived in 1929 to observe gorillas, but, untrained in field techniques, he obtained very little information on the animals themselves. During this period Robert Yerkes, the famous student of primate behavior, published an excellent book entitled *The Great Apes*, in which he brought together all available knowledge about the gorilla and other apes. I was impressed with how little was known then, and with how little information had been added by 1957. Most encounters with gorillas were brief, a fleeting glimpse, a moving shadow, a crashing branch as the animal fled. Could gorillas be studied, I wondered? A behavior study entails many uninterrupted hours of observation of undisturbed animals. At times I had the nagging doubt that I had started something I could not finish.

Then we heard of Walter Baumgartel, who in 1955 had bought a little hotel in Uganda at the foot of the Virunga Volcanoes. He was fascinated by the gorillas near his Travellers Rest Hotel and hoped to establish them as a tourist attraction. But he also realized the potential scientific value of the animals. He wrote to Dr. Leakey and Dr. Dart, two of the foremost students of apemen in the world, and asked their advice. As a result Miss Rosalie Osborn, a former secretary, attempted to study the gorillas between October, 1956, and January, 1957, and when she left, Miss Jill Donisthorpe, a journalist, continued the study until September, 1957. I awaited their reports with eagerness. When they arrived and I had read them, I was truly depressed. Here someone had made a serious attempt to study gorillas for a whole year, but the amount of concrete information about the behavior of the apes which these investigators obtained was minute. There were, as in previous accounts, descriptions of nests, of food remains seen along trails, of the roars of males in response to an intruder. Here and there were tantalizing glimpses of the ape's social life, but they were no more than glimpses, for the animals usually fled at the approach of the observer. My talks with scientists who had been in gorilla country did little to raise my spirits, for the general consensus of opinion was that I was attempting to do something highly difficult, if not impossible. Luckily, one evening I unburdened myself to Dr. C. R. Carpenter. No person knows primates better than he, and his pioneer field studies of howler monkeys in Panama and gibbons in Thailand are classics in their field.

"I think gorilla can be studied," he told me. "When I went to Asia to observe gibbons, everyone said that it could not be done. But if you develop your own techniques and find the right place to work, I am sure that you will be successful." His words cheered me immensely.

The reputed belligerence of the gorilla caused a certain amount of uneasiness and concern for the future, especially in my wife Kay. I realized, of course, that hunters had exaggerated their tales in order to be doubly acclaimed for their heroism in ridding the earth of such monsters, and that reliable authorities maintained that the gorilla is shy and retiring, seeking no trouble unless harassed. But the fact remained that Bingham and others, who were not actively hunting, were charged by males and had to shoot. Even Akeley, the gorilla's best friend, concluded a eulogy on the ape with the words: "I believe, however, that the white man who will allow a gorilla to get within ten feet of him without shooting is a plain darn fool. . . ." I refused to take a rifle or revolver into the field, feeling that firearms have no place in my kind of study. No animal attacks without good cause, except on rare occasions. My inclination is to give the charging animal all benefit of the doubt, hoping that it is merely bluffing. If the unusual happens and the charge is actually carried through, it is too late to shoot anyway. I had found this true with bears and other animals in Alaska, and I felt the same ideas applied to gorillas. Kay was not happy about my attitude, and I finally compromised by taking a little starting gun, such as is used to fire blanks at track meets. Since I never had to use it on gorillas, I was unable to determine its efficacy.

After months of discussion and meetings, the final plan of the expedition took shape. Doc would be the leader of the project and I his assistant. This arrangement pleased me greatly, for Doc has extensive experience in studying the ecology and behavior of mammals and birds, in addition to being an excellent teacher and congenial companion. He and his wife Jinny were to accompany the expedition for six months and then return to Wisconsin. My wife Kay and I intended to remain behind for an additional eighteen months to complete the study. In general, the purpose of the expedition was:

To study the distribution of the mountain gorilla, to note the various vegetation types which the ape frequents over the whole of its range, to ascertain similarities and differences in food habits, nesting habits, and other behavior from area to area, and, importantly, to search for a location where gorillas could be studied. We intended to devote the first six months of the project to this phase.

To observe intensively for at least one year the behavior of one population of gorillas.

To collect comparative data on monkeys and especially on other apes whenever possible.

Late in 1958, Doc and I received word that we could go. I was exhilarated. I must admit that, aside from the gorillas, Africa lured me. When imprisoned by today's cities, in the cement canyons and among the grating noises of machines, I find escape, at least mentally, by pouring over maps and reading travel books.

I project myself onto the hot sands of the Rub al Khali and onto the glaciers of the Mountains of the Moon. I had brought home books about Africa, stacks and stacks of them, until Kay almost refused to look at any more.

On February 1, 1959, two years after the topic of gorillas had first arisen, we left New York.

The purpose of this book is to present in popular form some of the results of our expedition. Most of it is naturally about gorillas, for these animals were the main incentive for our trip. At first I was somewhat diffident about adding yet another book on gorillas to the bookshelf. I have before me recent books by a hunter, a trapper, a hotel keeper, and a journalist. I myself have written a scientific monograph, *The Mountain Gorilla*, published in 1963 by the University of Chicago Press. But the books I have bought contain little information about the gorilla's habits when undisturbed, and my previous work is a compendium of facts, discussing the apes as subjects to be studied, not as acquaintances whose activities my wife and I discussed at the end of each day; I had no space to reveal the enjoyment I derived from roaming across grassy plains and uninhabited forests and climbing mist-shrouded mountains. In other words, this is to be a personal book about animals and the way they behave, of walks I have taken and sights that have pleased me. It is not an adventure book in the accepted sense of the word. Adventure implies hardships and accidents, which are usually the result of poor planning and carelessness. Our expedition accomplished, I think, what we set out to do without much trouble and, in retrospect, without great effort.

The support which the expedition received in the course of two years cannot be adequately acknowledged in a few words. Many of the persons who helped us are mentioned in the text, and I would here like to thank them all collectively. The National Science Foundation provided the funds, and a supplementary grant was received from the New York Zoölogical Society. This society, under the leadership of Dr. Fairfield Osborn, also acted as sponsor of the expedition, handling with great efficiency such mundane matters as correspondence and finances. In Africa we were fortunate to be under the patronage of the Institute of the Parks of the (Belgian) Congo and Makerere College in Uganda, both of which provided us with living quarters and logistic support. At home, the University of Wisconsin aided with facilities and equipment. Major credit for any success that may be attributed to the expedition belongs to Dr. Emlen. The entire program, from the initial application for funds to the completion of the project, was carried out under his able direction and guidance. He was instrumental in developing our field techniques and in gathering much information which he generously permitted me to use. I would also like to take this opportunity to thank Doc for all he has taught me directly and indirectly over the years, both in the field and in the laboratory, not only on an academic level but also on a personal plane. This book was written in 1962–63 while I was a Fellow at the Center for Advanced Study in the Behavioral Sciences at Stanford, California. Mrs. Hildegarde Teilhet helpfully typed the manuscript.

My thanks are due Holt, Rinehart and Winston, Inc., for permission to reprint lines from "Wilderness," by Carl Sandburg, on p. 152. The Ben Roth Agency, Inc., has given me permission to reprint "Am I satyr or man?" © *Punch*, London, on p. 223; unfortunately, I have not been able to identify the author of these lines.

CHAPTER 1

In the Realm of the Mountain Gorilla

The musty, somewhat sweet odor of gorilla hung in the air. Somewhere ahead and out of sight, a gorilla roared and roared again, uuua-uuua! *an explosive, half-screaming sound that shattered the stillness of the forest and made the hairs on my neck rise. I took a few steps and stopped, listened, and moved again. The only sound was the buzzing of insects. Far below me white clouds crept up the slopes and fingered into the canyons. Then another roar, but farther away. I continued over a ridge, down, and up again. Finally I saw them, on the opposite slope about two hundred feet away, some sitting on the ground, others in trees.*

There is a special feeling of elation in finally having reached a destination that has occupied the mind for months, a destination that is completely outside one's experience. The physical setting of Rumangabo is a delight, and during our stay there we never tired of looking out over the country below. In the mornings the air is wonderfully crisp and dew glistens on the foliage. The crimson blossoms of flame trees shine with unmatched brilliance. A slight breeze rustles the tall stands of elephant grass that spring up wherever the machetes of workmen fail to keep them in check. Far in the valley, toward the southeast, and on the slopes of the hills, huddle clusters of huts, smoke rising through the grass roofs. Most of the land is heavily cultivated, for the rich volcanic soil in this area supports one of the densest human populations in Africa. There are patches of bananas, small fields of beans and sorghum, and here and there stands of Australian wattle tree, planted for firewood because of its rapid growth. A few women hoe in the fields, and others balance large earthen jars on the head as they walk to the village wells. It is a peaceful scene, seemingly unchanged for centuries.

Civilization has, of course, intruded, and changes for better or worse have always occurred. Trucks rumble along the highway that passes by the base of the Rumangabo hill, and Africans dressed in tattered army surplus clothing pass on their way to work at the coffee plantations and on the roads. It is difficult to realize that the first European did not enter this region until 1894, within memory of the oldest living inhabitants, and less than seventy years ago, in 1898, the whole region was devastated by the cannibalistic Baleka tribe. Grogan, who walked through the area at that time, found that "every village had been burnt to the ground, and as I fled from the country I saw skeletons everywhere. . . . Mummies, skulls, limbs, putrefying carcasses washing to and fro in every limpid stream, marked the course of the fiendish horde." When he entered a hut, he found "a bunch of human entrails drying on a stick. A gnawed fore-arm, raw. A head with a spoon left sticking in the brains."

We had arrived in Rumangabo, the headquarters of Albert National Park, on February 14. We had decided to make Rumangabo our main base while we searched for a suitable area to conduct our gorilla study. The station, at an altitude of well over 5,000 feet, consists of a few buildings lying on top of a hill a little off the main road. The most imposing structure is the home of the commandant, Marc Micha, and his family. Micha, a short, compact man, is a former army officer like so many park officials in Africa. As we talked with him in his spacious living room and listened to his problems as administrator of the five Congo and Ruanda-Urundi parks, I could see that army training would be highly useful to him. Albert National Park alone employs over 250 Africans as guards against poachers and as guides for tourists. The discipline of the guards is maintained along army lines. On entering Rumangabo, we saw the guards armed with spears snap to attention, clicking the heels of their bare feet. They are smartly dressed in khaki shorts, jacket, and dark green fez. Throughout the day groups of guards march along in military formation, singing songs. And in the morning all are rousted out of bed to do push-ups and knee bends.

When we reached Rumangabo, the only other Europeans there besides the Micha family were two young bachelors, Rousseau and Bouckaert. Rousseau, lean and ramrod-straight, was in charge of the guards. His office as well as his home were by a small parade ground, and there we heard him barking the orders of the day. Bouckaert was head of maintenance, including the garage and generator.

Rumangabo had no tourist accommodations except the two guest houses we occupied. Each had a small bedroom, living room, and a primitive kitchen consisting of a tiny wood stove and a sink. We were provided with a servant too. Donati Endonyabo was a cheerfully inefficient soul who had been with the park for many years. He made our beds, washed the dishes, and spent much of his time blowing into the stove trying to get the fire going.

Beyond Rumangabo and the villages and fields loom the Virunga Volcanoes. Toward the southwest, across an expanse of forest which is part of Albert Park, rises Mt. Nyamuragira, 10,000 feet high and the lowest of the eight volcanoes. Its gentle contours belie its violent nature; every few years fiery rivers of lava descend from its slopes into the lowlands. Its neighbor, Mt. Nyiragongo, is over 1,000 feet higher, with steep walls and a flat summit that looks as if it had been sliced off with a giant knife. Mt. Nyiragongo is one of the few active volcanoes in the world which has a molten lava lake in its crater. In the evening of May 17, 1894, the German Count von Götzen, the first European to camp in the shadows of the Virunga Volcanoes, was in his tent when an African ran to him. "The sky is burning, sir!" he yelled, and looking toward the summit Von Götzen saw the glow from the lava lake of Mt. Nyiragongo. The name of the mountain means Mother of Gongo, Gongo being one of the most important spirits in the region. The Bahunde, a tribe that lives on the slopes of the rift escarpment, believe that the ghosts of the dead dwell in the depth of the mountain and stir the fires. When they fight among themselves, the earth shakes, the trees fall and point their roots toward the sky, and flaming water pours from the heights and cools to rock.

The most recent eruption in the area occurred in August, 1958. One afternoon Rousseau pointed out to us a mound of black lava near the base of Mt. Nyamuragira. Lava had suddenly poured from the forest floor and cut a swath through the trees. He told us of his visit to the new volcano, of the heat, and of the flames shooting far into the sky. Elephants had rapidly left the vicinity of the eruption, but the rodents remained, scurrying about within a few feet of the molten rock. Insects were attracted by the flames at night and perished by the thousands; bats chased the insects, dodging after them through the hail of rocks and ashes. Now the volcano was dead. Rousseau, with some pride, also told us: "The guards named the new volcano for me. His name is Kitsimbanyi. That means one who arrives suddenly." I thought this a good name, not only for the volcano but also as a description of Rousseau, whose habit of suddenly pouncing on lazy guards was already known to us.

Hundreds, perhaps thousands, of small lava plugs pock the terrain around the Virunga Volcanoes. Most are old, some densely overgrown with vegetation, others cultivated on the steep slopes. They attest to the fiery turmoil that persists beneath the earth's crust. At some future time, the villages and towns near the Virunga Volcanoes will surely be destroyed by lava flows like the most recent one, which headed toward the town of Rutshuru but stopped just short of reaching it.

Of the other volcanoes, no peak drew our eyes as often as Mt. Mikeno, which rises in rugged splendor to an altitude of 14,553 feet just south of Rumangabo. Once Mt. Mikeno was a volcanic cone, but over the years the slopes have eroded leaving only the central rocky core. Its bleak rock faces and deeply carved canyon give it a forcefulness of character which the other volcanoes seem to lack. Mt. Mikeno means "the naked one," and its pointed summit is almost devoid of vegetation. The huge mass

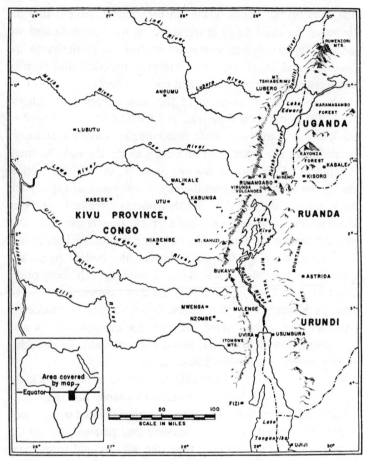

Regional map of the eastern Congo, western Uganda, and western Rwanda, showing the major physical features in the range of the mountain gorilla.

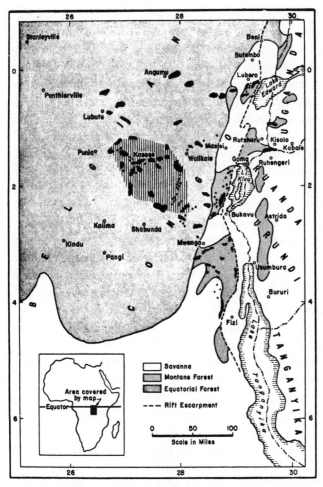

The distribution of the mountain gorilla with respect to vegetation types. Mountain rain forest and bamboo are lumped under montane forest.

The black areas indicate the location and approximate shape of isolated gorilla populations or gorilla concentrations in areas of continuous distribution. The small dots represent records of gorillas outside these areas. The hatching marks a central region of continuous but sparse distribution.

The Virunga Volcanoes are represented by the black area near the village of Kisolo; the Kayonza Forest lies just to the north of the Virunga Volcanoes.

of the mountain, changing its aspect every hour, now stark, now almost evaporated in the hazy air, is truly a gift to the spirit. Behind Mt. Mikeno and hidden by it rises the beautiful cone of Mt. Karisimbi to a height of 14,782 feet. Farther to the east is flat-topped Mt. Visoke. On clear days we could also see the serrated summit of Mt. Sabinio and the peaked top of Mt. Muhavura, but not Mt. Gahinga, which lies hidden between them.

The Virunga Volcanoes straddle the international boundaries of the Congo, Uganda, and Rwanda. The Congo and Rwanda portions are included in Albert National Park. It took us several days to straighten out not only the names of the peaks but also their spelling. The Virunga Volcanoes, for example, are also known as Birunga, Bufumbiro, and Mfumbira volcanoes. The word "virunga" is derived from "kirunga," a native term meaning "high isolated mountains that reach the clouds." Some of the early explorers brought in from the coast porters who referred to the volcanoes as Bufumbiro, the "mountains which cook." And European explorers created considerable confusion in naming the mountains by failing to understand the natives correctly. For instance, von Beringe discovered Mt. Visoke in 1899. Not realizing that when asked about a peak the natives give the name of the district in which it lies rather than the name of the mountain itself, he noted down the word "Visoke." Actually the name of the peak is Mago, and it lies in the district of Bisoko.

Before reaching Rumangabo we ordered a Volkswagen Kombi from a local dealer, and the park had the car waiting for us. Two days later, we drove toward Goma, the nearest town some thirty miles to the south, to open a bank account, obtain a car registration, buy provisions, and, importantly, check on our luggage which should have arrived long ago. Old lava flows covered with brush flanked the road on both sides for much of the way, and herds of goats foraged in the shrubbery. Here and there were villages of grass huts, and women, bowed almost to the earth, stumbled along under loads of wood. The road was narrow and torn up, and gangs of workmen pecked at the lava boulders with steel bars in desultory fashion. The road winds upward into the saddle between Mts. Nyiragongo and Mikeno and descends

The porters gather at Kibumba to haul our supplies into the mountains. The park guard (right) is sipping banana beer from a gourd.

Our hut at Kabara, at an altitude of 10,200 feet.

Dr. Emlen discusses the route of the gorillas with the guide Reuben Rwanzagire on the slopes of Mt. Muhavura.

A stand of tall bamboo in the Echuya Forest of Uganda. Similar stands exist on Mt. Tshiaberimu and in the Visoke-Sabinio saddle of the Virunga Volcanoes.

slowly to Lake Kivu and the town of Goma lying at its northern shore.

Lake Kivu was the last of the great lakes of the rift to be discovered. The explorer Speke heard of it in 1861 and named it Rusizi Lake after the river which connects it to Lake Tanganyika. In 1871 Stanley also placed the lake on his map, calling it Kivoe Lake, but he did not see it. Not until June 3, 1894, did Count von Götzen set eyes on its blue waters. Although sixty miles long and thirty miles wide, it is an intimate lake, nestling between escarpments that rise straight from the water's edge. Islands break the lake's expanse and peaceful coves and inlets interrupt the shore line.

Goma, meaning "drum" in Swahili, is the European center for this area. About half a mile of neat stores, painted in pleasant pastels, faces the main street. There are several hotels and banks, hardware stores and grocery stores, bakeries and open air cafés; along the side streets are garages and warehouses; and, at the outskirts, the native sector with houses built mostly of old tins hammered flat. Small African girls wander along the streets selling tomatoes and onions, and cripples beg for coins.

In the stores we were treated with the brusqueness and indifference usually reserved for tourists until we indicated that we would be around for some time. Immediately the mien of the tradesmen improved. Not being tourists, we were automatically classed as missionaries. Since one of the French words for "expedition" is "mission," I am sure that, with our far-from-fluent French, we sometimes conveyed the impression that we had come to convert the gorillas. We soon learned that to make any progress in such matters as permits and registrations we had to side-step minor officials and immediately try to see the top ones. Self-important but unwilling to make their own decisions, clerks shifted us from office to office until, several hours later, we were back where we had started. The fact that we were Americans did not bring out the most helpful responses anyway. The Leopoldville riots had occurred only a month before and now the United States was pressuring Belgium to speed independence for the Congo, an interference which the people resented.

Only about a mile from Goma lies the town of Kisenyi, in Rwanda. At the time of our visit both the Congo and Ruanda-Urundi were under Belgian control, and there was no custom post between the two countries when we drove that way. Kisenyi we found to be a delightful resort town with few stores, for nearby Goma is the business center. The main street is shaded by palms; many homes and small hotels, covered with masses of purple-blossomed bougainvillaea, lie along the shore of the lake. The waters of the lake are delightfully cool, and a sandy beach makes bathing a pleasure.

Goma and Kisenyi were tourist resorts when we saw them, a far cry from the early part of this century when they were the site of a border dispute between the Germans and Belgians. If one nation constructed a military post, the other immediately built one nearby to check on activity. Wollaston published this description of the towns as they were in 1908 in his book *From Ruwenzori to the Congo.*

Close to the place where we first came down to the north shore of the lake was the temporary Congo post of Ngoma. We found there a small garrison and one Belgian officer living in miserable grass huts. . . . A couple of miles along the shore, near the north-east corner of the lake was the German post of Kissegnies. The Germans had settled upon a beautifully fertile piece of land, which they had cleared and cultivated; they had made roads and planted trees. The brick-built house of the commanding officer was a miracle of solid walls and coolness inside. . . .

The road we followed westward from Goma to the village of Sake, at the base of the rift escarpment, led across a series of spectacular lava flows. Eruptions in the area were recorded in 1904, 1912, 1938, 1948, 1950, 1951, 1954, 1957, and 1958, most of them near the slopes of Mt. Nyamuragira. Perhaps the most spectacular eruption was witnessed by Sir Alfred Sharpe in December, 1912.

All the country along the shore of Kabino was buried in black ash, the crops were destroyed, banana trees fallen, native huts partly buried or crushed flat; the incessant rain coming from the black cloud overhead caused land slips which carried crops and houses into the lake. . . . For miles in every direction was black; there was not a green leaf or blade of grass to be seen. We found many birds and small mammals killed by falling stones, some of which measured 2 inches in diameter. We did not sleep that night: we heard several sharp earthquake shocks, hurricanes of wind frequently occurred with appalling lightning, our tents were nearly blown away, and for two hours a heavy fall of ash and stones threatened to bury our small camp. The roar from the volcano was incessant — a steady deafening roar — and the whole country below us was lit up by the column of fire lava, and red-hot material, which was shot up many thousands of feet. . . . The whole of the water at the north end of Kivu was hot by this time, and many thousand fish were floating dead. . . .

Many hundreds of natives were killed, mostly owing to their refusal to leave their villages and seek shelter elsewhere. . . . Some idea of the fierceness of this outbreak while it lasted may be gathered from the fact that at the post of Walikali, in the Congo forest 100 miles to the west, ashes fell heavily for two days, while the eruption was heard at Beni 140 miles to the north. . . ." (Geographical Journal, XLVII, 1916)

The ages of many of these flows are known, and I found it interesting to note the rapidity with which plant life took hold upon the sterile rock. In 1904 a river of lava poured into Lake Kivu, but soon lichens, the first colonizers, covered the rock with a hoary coat. Ferns found footholds, and seeds sprouted from dung which wandering elephants had dropped. Organic matter accumulated in the cracks and fissures, and various herbs found homes there. Over the years, dense stands of saplings and tangles of brush appeared, and now trees over thirty feet high shade the once barren lava.

André Meyer, a geologist living in Goma, has done considerable research on the volcanoes, and in reply to our questions sketched the broad outlines of the recent geological history of the area. Long ago the valley was peaceful. Forests grew on the escarpment and a river meandered lazily toward the north. But about half a million years ago intense volcanic activity began, during which two volcanoes, Mts. Mikeno and Sabinio, erupted from the valley floor. The activity subsided somewhat until new forces were unleashed within the last one hundred thousand years when Mts. Karisimbi, Visoke, Gahinga, and Muhavura emerged. The final major convulsion of the earth's crust occurred about twenty thousand years ago and created Mts. Nyiragongo and Nyamuragira. As lava flow upon lava flow spread across the valley it created a dam eighteen hundred feet thick and fifteen miles wide. Each flow averages about six hundred feet in width, and drilling at one spot to a depth of fifty feet revealed eight separate layers, attesting to the innumerable flows that made up the dam. The river which once flowed toward Lake Edward found its way barred. Slowly the waters rose in the valley, creating Lake Kivu. It took four thousand years to fill the valley. Sea shells high on the walls of the escarpment indicate that the waters rose almost four hundred feet above the present lake level before finding an outlet in the Ruzizi River to the south.

We eagerly absorbed all the new sights around us. Every day was filled with new discoveries and experiences. Doc and I drove to Bukavu, the capital of the Kivu province at the south end of Lake Kivu, to pick up our luggage; we all visited the plains of Albert Park to watch the herds of elephants, topi and kob antelope, and buffalo; we went to Kampala, the largest city in Uganda, to buy supplies and pay our respects to Dr. A. Galloway of Makerere College, one of the local sponsors of our expedition; we prepared for our future gorilla study in Uganda by asking the advice of Major Bruce Kinloch, John Mills, and other members of the Uganda game department. In this way we laid the essential groundwork for the coming months by becoming acquainted with the area and the persons who were interested in furthering our study. But I was becoming impatient. We had left the United States over a month before, and I still had not seen a gorilla. Fortunately, Doc lacks my impulsiveness, and from previous expeditions to Africa and Central America he had learned the wisdom of having the proper support before proceeding with work.

Our first extended visit to the Virunga Volcanoes was arranged for March 6. Unfortunately, the Institute of the Congo Parks in

Brussels, which, not excepting Micha, is the only organization that can give permission to visit the gorilla country in the region of the volcanoes, limited us to one week in the area. We were understandably unhappy that we were so restricted in the time we were alloted, but we hoped for the best. Our spirits were considerably raised by Dr. Curry-Lindahl, a Swedish zoölogist, who had just returned from the volcanoes. He had encountered gorillas several times and once watched a group of them for half an hour.

On the appointed morning, with Rousseau, who had arranged the trip and was to accompany us, we all drove to Kibumba, a village at the base of Mt. Mikeno, where a swarm of porters was waiting. Jinny and Kay came along to see us off. We unloaded boxes, baskets, duffels, and our other gear, all neatly tied and the weight distributed so that each load weighed about thirty pounds. The porters, hired at 30 francs (60 cents) a day, crowded around, judging the loads, shouting and pushing. Children ran giggling around the periphery, and the three park guards dashed back and forth assigning a man to each load. The porters ripped handfuls of long grass from the wayside and twisted and tied it into a small ring which they then placed on the head as a cushion for the load. Finally we were off, sixteen porters and three guards in a single file, with us tagging behind.

The method of transporting loads in Central Africa has not changed since Stanley's time, and I felt out of my century as I walked behind the scraggly line of porters, bundles balanced on their heads as they moved along narrow paths between fields of beans and maize. Another heritage of the early explorers seems to be a tendency for Europeans traveling afield to take with them vast numbers of Africans. In Alaska I had become accustomed to carrying on my back enough supplies for two or more weeks, and I wondered what we had taken along to require sixteen porters. My curiosity was satisfied later when the houseboy Philippe, whom we had hired for the journey, unpacked Rousseau's gear, including a heavy mattress and sheets, folding tables and chairs, a complete kitchen service including china bowls, pitchers, and plates, and a huge basket containing fresh fruits and meats. We had all the comforts of home. I remembered reading about a 1954 expedition to the volcanoes to study the gorillas. One hundred and twenty porters were taken for a one-week trip. Of course they saw no gorillas. A talking and laughing horde not only destroys the tranquillity of the forest but chases all the animals over the next hill. One evening at a dinner party I imprudently expressed surprise when someone related that he had recently taken eighteen porters for a three-day trip.

"You will learn," the person replied with a smile. "Africa is not America. No one carries his own things here."

Later, for trips of a week or less, I took two or three men with me, one to lead the way and the others to keep him company. Loads were divided evenly among all of us. For trips such as this one to the volcanoes, I continued to use as many porters as necessary to haul our equipment to base camp, but then discharged them immediately, keeping only two in camp.

We followed a maze of trails which in the ensuing months I learned to know very well. The path led through a village with grass huts like hay stacks standing in rows. Goats and children scampered between the storage bins that stood opposite the huts, and women were drying sorghum seed on reed mats in the morning sun. Pied wagtails flitted around the huts, bringing, the natives believe, good luck to their households. After a walk of half an hour we reached the edge of the forest. The porters stopped to adjust their loads, and then we entered the home of the gorilla.

For the first half a mile the forest was uninspiringly scrubby. *Mimulopsis arborescens*, a tree-like shrub with large toothed

leaves and straggling branches, crowded the edge of the trail. Where the forest had recently been disturbed grew hedges of *Acanthus pubescens*, with magenta flowers and spiny leaves that made the barefooted porters step gingerly. Towering above the undergrowth were scattered *Neoboutonia macrocalyx* trees with smooth grey trunks and broad light-green leaves that turned to shimmering silver when a breeze stirred them. All the brush had recently been cut near our footpath, making a swath over fifteen feet wide. Why, I asked Rousseau, was this unnecessary cutting done in an area that was rarely visited and supposedly protected?

"It is King Leopold," he answered with a shrug. "He wishes to climb Mt. Mikeno soon. So we make it easy for him."

Our surroundings changed abruptly when we entered the zone of bamboo, a grass which grows in the highlands between the altitudes of eight and ten thousand feet. The canopy closed in over our heads as we walked through a green bower. Rays of sunlight played among the ranks of stems around us. Glancing up, our eyes met only the translucent screen of bamboo leaves, eerily green — it was like looking toward the surface of the water from the bottom of the sea. Our trail was now only a buffalo and elephant path winding along the edge of the Kanyamagufa Canyon that slashes the slopes of Mt. Mikeno. According to native legend a great fight occurred here long ago and the men who were killed were thrown into the chasm, afterward named Kanyama-gufa, "the place of bones." The guards now walked ahead and cleared the trail with swift strokes of machetes. I marveled at our nimble porters as they dodged under branches, scrambled over downed trees, and jumped across elephants tracks, never upsetting their loads and all the while shouting back and forth.

At first the gradient of the trail was low, but it soon steepened. Talking ceased as we puffed upward, and our shirts were dark with sweat. After about two hours we reached Rweru, a small meadow surrounded by a fringe of trees on top of a knoll. This was a traditional resting place of the porters and the approximate halfway point to the saddle between Mts. Mikeno and Karisimbi, our destination. The men adjusted their loads again, some lay in the grass smoking cigarettes, a few ate cooked beans, and others wandered over the edge of the hill to relieve themselves.

From Rweru the Kanyamagufa Canyon turns in a northerly direction, and so did our trail. The walls of the canyon were perpendicular; several hundred feet below I saw the rocky stream bed. For how many centuries had our trail existed? Generation after generation of buffalo had followed the edge of the canyon, kicking the fragile soil with their sharp hooves. Earlier visitors to these mountains describe the same route, the same canyon, the same trail we followed. Prince Wilhelm of Sweden, Carl Akeley, King Albert, Sir Julian Huxley — all had labored ahead of us. The bamboo suddenly ceased, and we entered *Hagenia abyssinica* woodland. We were in an enchanting parkland, the views suddenly wide, for little brush grew here and the trees were scattered. To our left rose the massive citadel of Mt. Mikeno, and to our right Mt. Karisimbi, sloping softly and gently toward the clouds.

The *Hagenia* became my favorite tree. Only about fifty to sixty feet high, it lacks the remote impersonality of the forest giants whose branchless trunks rise over a hundred feet into the air. Its long pinnate leaves cluster in an umbrella-shaped crown, and pendents of reddish flowers hang between the sparse foliage. The trunk is stocky, up to eight feet in diameter, and near the ground massive branches flare horizontally from the trunk before rearing skyward. The ochre wood is not durable, and many trees are hollow, ideal retreats for whatever forest sprites inhabit this woodland. The bark flakes raggedly, soft cushions of moss pad the limbs, ferns hang beneath, and lichens cover the branches, giving the tree the appearance of a kindly, unkempt old man.

Doc and I were not in condition for climbing, and our legs ached and our breath came in gasps at the 10,000-foot level. The porters suddenly shouted gaily, and soon the path widened to a small meadow with a cabin at its edge. We had reached Kabara, "the place of rest." I collapsed in the cool grass, and the porters, still spry after carrying a load for five hours, laughed at me.

The cabin, of roughly hewn boards and covered with a tin roof, had been standing there about twenty-five years. There were three large rooms and two sheds strung out in a line, a highly unimaginative design for a mountain hut, for it proved impossible to keep warm. Inside were two tables, three chairs, a tiny wood stove, two bedsteads, and some ragged grass mats covering the walls of one room. The three windows — one to each room — admitted little light for the panes were barely translucent. At the corners of the hut stood rusty iron barrels to catch the rain water as it drained from the roof. A small outhouse, door torn off the hinges, stood behind the cabin.

In one corner of the meadow, partially shaded by trees, lay a grave with a large flat stone bearing the simple inscription: *Carl Akeley, Nov. 17, 1926.* Surrounding the stone was a low wall of twisted lava rock. Carl Akeley's wife, who had been with her husband when he died, had returned in 1947 to build this wall, now overgrown with the yellow blossoms of sedum. Doc and I stood before the grave, feeling ourselves challenged by Akeley's unfinished study of the gorillas. Looking from his grave across the green meadow dotted with delicate flowers and upward across the sweep of forest to the summit of Mt. Karisimbi, we knew why Akeley had considered this spot one of the loveliest on earth. At the other end of the meadow, about four hundred feet away, was a shallow pool, its surface cloudy and its banks churned by the hooves of many buffalo. And all around us was the forest.

It seemed I had barely gone to sleep that night when I heard the Africans talking quietly as though at a great distance. I crawled from my sleeping bag, shivering as I dressed. Outside, the sky was clear and the world into which I stepped was silent. The porters and guards lay wrapped in their coats and blankets by a fire in one of the sheds. The air had the crispness of northern climes. Slowly, as the sun rose over the distant slopes, the trees turned a delicate gold and the grass sparkled with dew. A red forest duiker emerged from the bushes at the far end of the meadow, and the sun turned its coat red as copper. This delicate antelope, no larger than a setter, grazed shyly, looking in my direction every few seconds, and as I watched it jumped with a series of high bounds into the undergrowth.

After breakfast, Doc and I decided that to find gorillas we should explore independently rather than searching the forest together. I clambered alone up the steep slope of Mt. Mikeno behind the cabin, then parallel to the slope, hoping to come upon a fresh trail. The undergrowth of thistles, nettles, wild celery, and other herbs was a good six feet high. My feet caught on hidden roots and slipped on succulent herbs; stinging nettles slapped me across the face, leaving red welts. Once a duiker bounded up at my feet and with a sharp *pshi-pshi*, perhaps made by blowing air through its nose, scurried away, leaving behind a strong musty odor. Involuntarily I jumped back a step, and my heart beat unevenly.

It is an exhilarating experience to wander alone through unknown forests when everything is still new and mysterious. The senses sharpen, bringing into quick focus all that is seen and heard. I knew that there were leopards on these slopes, and the trails of the black buffalo crisscrossed the area. Both animals have the reputation of being unpredictable, and I was watchful. Each creature has its own distance at which it will take flight from an intruder, and each will allow itself to be approached only so far before defending itself. It behooves man to learn the responses of each species; until he has done so, and is familiar with the sights, sounds, and smells around him, there is an element of danger. Even so, danger that is understood does not detract from but rather enhances the pleasure of tracking.

The first signs of gorillas I came across were three nests on the ground. Branches from the undergrowth had been pulled toward the center creating crude platforms on which the animals had slept for the night. But the broken vegetation was wilted and the dung at the edge of the nests was covered with small red fungi, indicating that the site was old. I was heartened by my find, and, after examining the nests, I pushed on. Neither Doc nor I had much luck that day. We found nothing but old sites.

The following morning, Rousseau wanted to check the Rukumi rest house, which lies another thousand feet higher, on the slopes of Mt. Karisimbi. We joined him, and I took my sleeping bag in the event that I decided to remain there for the night. For an hour we climbed upward through the *Hagenia* woodland. Then these trees grew sparser and were replaced by *Hypericum lanceolatum*, a bushy tree with small, lanceolate leaves and striking yellow blossoms. Abruptly the terrain flattened, the forest ceased, and we stood at the edge of a huge meadow with the summit cone of Mt. Karisimbi towering three thousand feet above us. The dark branches and trunks of the trees growing here and there in the meadow were in vivid contrast to the yellow of the grass. The small bell-shaped flowers on the branches of these trees I recognized as heather —not the small shrub most of us know but a heather tree over thirty feet high. Among the heather grew wild blackberries, large and delicious, on which we feasted until our fingers and mouths were purple-stained. The rest house stood against the base of the mountain with a magnificent view across the meadow, past the black silhouettes of the heather to the bleak summit of Mt. Mikeno.

The slopes of Mt. Karisimbi beckoned, and I knew that I must climb the peak. Doc came a short distance to examine the most unusual forest we had ever seen. Tree groundsels or giant senecios, weird and gnarled plants with an other-worldly look, were scattered over the open slopes to an altitude of 13,500 feet. In the temperate zones of the world, senecios are insignificant weeds, but here in the cold and humid mountains they are giants over twenty feet high, with thick stems and flowering heads a foot or more long. The large ovate leaves are clustered at the apex of the branches and they glisten as if polished. The only other tall plants in this curious forest were giant lobelias, consisting of a single stem, a cluster of narrow leaves, and pointing skyward like a candle a long flowering head covered with tiny purplish blossoms.

I continued alone, plodding upward along buffalo trails, reaching for the summit that was now lost in the clouds. After an hour the senecios were stunted and the tracks of buffalo were fewer. The whole slope was covered with a springy mat of

Alchemilla, at places over a foot deep. The altitude affected me now. My body felt heavy. Fifty steps, stop and breathe; another fifty steps, stop and breathe. Sometimes my foot caught in the vegetation, and I pitched forward. I lay on the ground, arms spread, face pressed against the cool soil, until I felt ready to continue. Bare patches of earth and outcrops of lava were signs that I was nearing the summit. Heavy, clammy clouds pressed down on me, increasing my feeling of solitude. I came across a single buffalo track, a dark line in the soil, snaking upward into the fog and I wondered what possesses man and beast to seek these barren heights? Dr. James Chapin, who devoted many years to the study of Congo birds, found the skeleton of an owl monkey (*Cercopithecus hamlyni*) on the summit of Mt. Karisimbi, miles from its forest home. And recently I came upon a fascinating account of a pack of wild dogs seen on the glaciers of Mt. Kilimanjaro at nearly twenty thousand feet. Perhaps man is not the only animal that climbs a mountain merely because it is there.

Some two and a half hours after leaving the Rukumi meadow, I stood on the summit, marked by a shallow crater, a rain gauge, and an assortment of boards, cans, and other rubbish discarded by the many expeditions that had reached the top since the first climbers in 1903.

I waited twenty minutes for the clouds to disperse, but when it started to hail I hurried down the slope. I remembered the tragedy of the geologist Kirschstein who was surprised by a snow storm while descending in Februray, 1907. His porters, in snow and bitter cold for the first time in their lives, lay down and refused to get up again. They moaned, "It is the decree of the gods — we must die." Twenty porters died in that storm, and Kirschstein, in his effort to drag the half-frozen Africans to shelter, caught pneumonia and lay unconscious for two days.

Numerous deep ravines radiate from the summit like the spokes of a wheel. Unable to see more than fifty feet ahead, I did not follow the same path I used in climbing up. Far down the mountain, I took a compass bearing and knew that I had strayed. I climbed in and out of ravines as I crossed the slope. The rock was wet and slippery, and the stands of senecios pre-

sented a chaotic mass of brittle stems that snapped underfoot, pitching me into the sodden moss that carpeted the ground. It was an eerie shadow-world, full of grotesque shapes, monsters that appeared for a moment only to vanish again. Suddenly a buffalo rose out of the swirling fog, standing there black and ominous, head slightly lowered to reveal the sweeping curve of its horns. I stopped until it disappeared soundlessly like a phantom. Only later, far away, I heard the breaking of branches.

When I reached the rest house only a guard was waiting for me; the others had returned to Kabara. I stripped off my wet clothes and, wrapped in a blanket, squatted by the fire until darkness fell.

For the next two days we continued our search for gorillas. We found some fresh feeding sites, and Doc once thought he heard a gorilla in the distance. On the third day, far down along the Kanyamagufa Canyon, I heard a sound that electrified me — a rapid *pok-pok-pok*, the sound of a gorilla pounding its chest. I followed the edge of the canyon until I found a game trail that crossed it. Carefully I scouted along the slope where I expected the animal to be. But I had no luck. Only later did I learn that there is an almost ventriloqual quality in the sound of chest-beating that makes distance very difficult to judge.

When Doc and I returned to the canyon in the morning we were greeted by the same noise. Evidently a gorilla had spotted us. I climbed into the crown of a tree to look over the shrubs that obscured our view, and Doc circled up the slope. Suddenly, as he told me later, the undergrowth swayed forty feet ahead, and Doc heard the soft grumbling sound of contented animals. Unaware of him, the gorillas approached to within thirty feet.

Two black, shaggy heads peered for ten seconds from the vegetation. Uncertain of how to react, Doc raised his arms. The animals screamed and walked away. We both examined the swath of freshly trampled vegetation and the torn remnants of wild celery and nettles on which the gorillas had been feeding. While Doc took notes on the spoor, I followed the trail. The musty, somewhat sweet odor of gorilla hung in the air. Somewhere ahead and out of sight, a gorilla roared and roared again, *uuua-uuua!* an explosive, half-screaming sound that shattered the stillness of the forest and made the hairs on my neck rise. I took a few steps and stopped, listened, and moved again. The only sound was the buzzing of insects. Far below me white clouds crept up the slopes and fingered into the canyons. Then another roar, but farther away. I continued over a ridge, down, and up again. Finally I saw them, on the opposite slope about two hundred feet away, some sitting on the ground, others in trees.

An adult male, easily recognizable by his huge size and grey back, sat among the herbs and vines. He watched me intently and then roared. Beside him sat a juvenile perhaps four years old. Three females, fat and placid, with sagging breasts and long nipples, squatted near the male, and up in the fork of a tree crouched a female with a small infant clinging to the hair on her shoulders. A few other animals moved around in the dense vegetation. Accustomed to the drab gorillas in zoos, with their pelage lusterless and scuffed by the cement floors of their cages, I was little prepared for the beauty of the beasts before me. Their hair was not merely black, but a shining blue-black, and their black faces shone as if polished.

We sat watching each other. The large male, more than the others, held my attention. He rose repeatedly on his short, bowed legs to his full height, about six feet, whipped his arms up to beat a rapid tattoo on his bare chest, and sat down again. He was the most magnificent animal I had ever seen. His brow ridges overhung his eyes, and the crest on his crown resembled a hairy miter; his mouth when he roared was cavernous, and the large canine teeth were covered with black tartar. He lay on the slope, propped on his huge shaggy arms, and the muscles of his broad shoulders and silver back rippled. He gave an impression of dignity and restrained power, of absolute certainty in his majestic appearance. I felt a desire to communicate with him, to let him know by some small gesture that I intended no harm, that I wished only to be near him. Never before had I had this feeling on meeting an animal. As we watched each other across the valley, I wondered if he recognized the kinship that bound us.

After a while the roars of the male became less frequent, and the other members of the group scattered slowly. Some climbed ponderously into shrubby trees and fed on the vines that draped from the branches; others reclined on the ground, either on the back or on the side, lazily reaching out every so often to pluck a leaf. They still kept their eyes on me, but I was amazed at their lack of excitement.

"George," called Doc. "George!"

At the sound all the gorillas rose and disappeared silently at a fast walk. Doc had become concerned for my safety after first hearing the many roars and finally the prolonged silence. We ate our lunch — crackers, cheese, and chocolate — before we checked the site where I had watched them. Their trail angled upward across a valley and up another ridge, where we found the group again two hours later. A female sat on a mound, her infant beside her. The male, ever alert, roared when he spotted us and stalked back and forth in the usual posture of gorillas — feet flat on the ground and the upper part of the body supported on the knuckles of the hands. When he approached the female on the mound, she moved rapidly to one side, and he claimed her place. As before, the group settled down and seemingly paid us scant attention. The male, who must have weighed about four hundred pounds, rested on the mound looking out over the mountains and the plains, truly the master of his domain. A female holding an infant gently to her chest walked to his side.

"It must have just been born," I whispered to Doc. "It's still wet." And he nodded in agreement.

The female leaned heavily against the side of the male. Her hairy arm almost obscured her spidery offspring, whose hairless arms and legs waved about in unoriented fashion. The male leaned over and with one hand fondled the infant. For two hours, enthralled, we watched this family scene. But the way home was long, and reluctantly we left the animals, but not before we spotted another gorilla far uphill and barely visible. Was it another group or a single animal? With buoyant steps we moved down the slope. We only hoped that in our presence the gorillas would always be as tranquil as today. Perhaps, we feared, they had merely been loath to move because of the imminent or recent birth of the infant.

But something else took our minds off the apes. Just ahead and close to the trail that crossed the Kanyamagufa Canyon an elephant wheezed, and then another. We stopped and strained our ears, trying to locate the gray forms gliding softly through the bamboo and the brush around us. It was my first close contact with elephants, and I was nervous. I clambered up a tree with much noise as the dry branches snapped under my weight. It was embarrassing, for after all my scrambling I was still no higher than the elephant's back. Doc motioned me to come down, and we hurried in and out of the canyon, hoping that no one would bar our way.

On the following morning we returned to the ridge where we had seen the gorillas. The animals had descended into the valley and were now feeding leisurely. It struck us immediately that there were now more gorillas than yesterday. We counted, re-counted, and agreed there were twenty-two: four adult or silver-backed males, one young or blackbacked male, eight females, three juveniles, five infants, and one medium-sized gorilla of whose sex we were uncertain. Did two separate groups join, or had we seen only part of the group the day before? The big males roared and slapped their chests, but, as on the previous day, they seemed little concerned as they rested beneath the canopy of trees. As we watched, most of them lay down and went to sleep. One female sat with a large infant in her arms. Another small, woolly infant left its mother and bumbled over to the first female. Briefly she cuddled both youngsters to her chest. A male rose, casually ambled by a sitting female, suddenly grabbed her by the leg, yanked her two to three feet down the slope, then cantered off. It was a wonderful feeling to sit near these animals and to record their actions as no one had ever done before. We had the chance to observe significant and characteristic incidents, but we knew that to explain what we were seeing — and even to predict what might occur in a particular circumstance — would take many, many hours of observation.

After resting for some three hours, the animals spread over the hillside to feed. An infant ran along, put some leaves into its mouth, and spat them out. A juvenile, perhaps three or four years old, held to the end of a three-foot log with both hands. It bit off the rotten bark and appeared to lick a whitish fluid from the wood. Another juvenile came up and, having wiped the log with one finger, licked it. Soon all the gorillas were actively moving across the slope, feeding as they went. Their movements were restrained and rather phlegmatic; only the youngsters behaved in exuberant fashion. One infant dashed along, pounced on the back of another infant, and both disappeared rolling over and over into the undergrowth. After five hours with the gorillas we returned home.

A guard and I tried to find the group again the following day, but we had no luck. I realized how much I had to learn about the ways of the forest and about tracking animals over long distances. The park had no teachers, for the local Bantus were all agriculturalists; they avoided the forest and the wild animals and evil spirits it contained. I knew that before I could study gorillas successfully I would have to teach myself to recognize the age of spoor, the number of animals involved, and the direction they had taken. The forest was vast, the animals few, and I could not depend on luck in finding gorillas as we had done so far.

Luck, however, was still with us, for on the way home, at a place where the Kanyamagufa Canyon veers sharply toward the upper slopes of Mt. Mikeno, we came across a fresh trail. Early the following morning, when Doc and I had barely proceeded three hundred feet along this trail, the brush crackled ahead. A black arm reached from the undergrowth, pulled a strand of a vine from a branch, and disappeared from view. We stepped behind the bole of a tree and peered at the feeding gorillas about one hundred feet away. Without warning, a female with infant walked toward our tree, a large male behind her. I nudged Doc and quietly climbed up on a branch without being seen by the animals. The female stopped some thirty feet from us and sat quietly with her large infant beside her. Once the infant glanced up at me, then stared intently for fifteen seconds without giving the alarm. But when the female inadvertently looked in my direction, her relaxed gaze hardened as she saw me. She grabbed her young with one arm, pulling it to her, and with the same motion rushed away, emitting a high-pitched scream. The male answered with a roar and looked around, and Doc, having failed to interpret the purpose of my nudge, was surprised to see me in the branches above him. The members of the group assembled around the male after a moment of tense alertness. The animals were still within about one hundred feet of us, and we wondered what would happen. To our relief, one face after another turned toward us in a quiet, quizzical stare as curiosity replaced alarm. They craned their necks, and two juveniles climbed into the surrounding trees to obtain a better view. One juvenile with a mischievous look on its face beat its chest, then quickly ducked into the vegetation, only to peer furtively through the screen of weeds as if to judge the effect of its commotion. Slowly the animals dispersed and went about their daily routine. I particularly remember one female who left the deep shade and settled herself at the base of a tree in a shaft of sunlight. She stretched her short legs in front of her and dangled her arms loosely at her sides. Her face was old and kind and creased by many wrinkles. She seemed utterly at peace and relaxed as she basked in the morning sun.

After three hours of continuous observation we regretfully left the group. It was March 15, and we were due to depart from Kabara that day. The porters, most of whom had returned to their homes the day after we reached camp, came to fetch us at noon. We descended the mountain sad that this journey had ended and pleased that it had been a success. The dream which had inspired and plagued us through many months of planning and preparation had come true. To approach and observe gorillas at short range had turned out to be possible. We had caught such fascinating glimpses of the gorilla's home life that I knew that someday, after we had finished the general survey, I would have to return to Kabara with Kay to answer the many questions that filled my mind.

CHAPTER 2

The Impenetrable Forest

Slipping and sliding, we descended into a valley, into another world. The Batwa cut a path with their machetes, but at times we simply crawled several feet above ground over the swaying mass of vegetation. Thorns tore at us, and vines tugged at our feet. There were stands of tree ferns with black stems and fronds that flowed gracefully outward from the apex. The light in these groves of ferns was translucent green, subterranean in quality. Spines covered the stems of the ferns and often, when we stumbled, our hands automatically reached for a stem for support. Roaches scurried about. The atmosphere was primeval, paleozoic, as when the world was ruled by tree ferns and amphibians. Small streams sought their way down most of the valleys, and from these we drank deeply and cooled our hot faces.

The most pleasing way to approach the Virunga Volcanoes is from the north, through the Kigezi district of Uganda. The narrow road winds among rugged hills, most of which are heavily cultivated along contour lines. Clusters of grass huts with peaked roofs cling to the slopes. Little boys dressed in nothing but ragged shirts tend herds of checkered goats by the wayside. And back and forth from village to village, and across the valleys, the men and women call to each other, imparting the news of the day with a sing-song voice that echoes through the hills. In the valleys are streams and swamps and sometimes lakes to which the women descend daily from the high slopes to fetch water. Cultivation gives way to bamboo forest near the summits of some hills, and suddenly, at an altitude of eight thousand feet, you emerge on the crest of the Kanaba pass.

Far below lies the floor of the rift, and the whole chain of the Virunga Volcanoes is spread out in a vast panorama. Mt. Muhavura, a perfect cone with a fringe of forest along its lower slopes, is 13,540 feet high. Once this mountain served as a beacon, guiding the natives home from their distant journeys, and thus received the name "guhavura," the guide. Beside Mt. Muhavura and dwarfed by it is flat-topped Mt. Gahinga. And, barely visible in the dry-season haze, is the serrated summit of Mt. Sabinio, "the old man with large teeth," nearly twelve thousand feet high and the last of the three volcanoes which lie partially in Uganda. Near the base of Mt. Muhavura nestles the village of Kisoro. There are a few shops, run by Indians, who are the only traders in Uganda. The stores carry tinned goods, bolts of cloth, cooking pots, lanterns, fresh produce, kerosene, and many other items, piled here and there and hanging from the ceiling. A Uganda custom post is in Kisoro too, for here the roads part to nearby Rwanda and Congo.

Kisoro is a drowsy little hamlet with little to change the even rhythm of its life. Once the main road to the Congo passed through here, but a new one was built to circumvent the mountains. In the fields the men and women pry the boulders of lava from the ground and plant potatoes, maize, and beans in the fertile but shallow soil. The Indians remain in the dark interior of their stores, a world apart, dreaming of the day when they can return to their homeland as rich men. Only once a week, on market day, does the tempo of life quicken. The natives from all the surrounding villages congregate at the edge of town and bring their wares. There are bamboo baskets, woven from strips of pliable young stems and chinked with cow dung; there are mats of reeds, earthen pots, live chickens, sick sheep with hanging heads and draining eyes, and bowls of many-colored beans. The women are dressed in all their finery. The older ones still retain their covering of goat hides, and their ankles and wrists are heavily laden with innumerable coils of copper wire, heavily tarnished, the number of which was once a sign of wealth. But the younger women are wrapped in colorful printed cloth, usually with an infant slung on the back, only its brown head peering out. Children playing tag, run shrieking among the baskets, and men shout at each other in disagreement over a bargain. It is a happy crowd, for once the dreariness of life is forgotten, the spirit revives.

Kisoro is known throughout the world for its Travellers Rest Hotel. It is a small hotel — a main lodge, surrounded by gay beds of flowers, and three round huts, called rondavels, off to one side. There is no running water, and dinner is served by lantern-light. But it is the only place in Central Africa from which the average visitor can in one day enter gorilla country with a fair chance of seeing the animals and their nests.

To Kay and myself the Travellers Rest Hotel was our second home. After a few months in the Congo, tired of the cool reserve of the Belgians, we would sometimes drive to Kisoro.

"Nooo! How are you?" The hotel-owner, Walter Baumgartel, greeted us with outstretched arms, his eyes twinkling, a happy smile on his face; and we were content. Walter, who has wandered over the world for many years, loves the mountains that tower near his hotel, and above all the gorillas.

After our visit to Kabara, and at the invitation of Walter, we moved to Kisoro to become familiar with the gorillas in this region of the Virunga Volcanoes. For our first trip Walter promised to lend us his three guides, who in Central Africa have no equal as gorilla trackers. Reuben Rwanzagire, the head guide, is unusual among Bantus in that he seems to derive pleasure from roaming the mountains. His frame is spare and tall and his face ascetic. Although his high forehead and gently curving nose give him the slightly hamitic look of a Watutsi, those lean giants in nearby Rwanda, he claims to be pure Bahutu. Ntirubabarira and Bagirubgira, the other two guides, seemed rather withdrawn to me. It is never easy to get to know the natives. They usually remain in a world of their own, and most attempts to draw them out are met with silent rebuff. In this Reuben differed. His English was limited to "yes" and "thank you" and his Swahili, the *lingua franca* of eastern Africa, was not very good — though far superior to the hundred or so words we had learned — but he had a genuine interest in conversing.

On March 25 we met the porters at the base of the mountains, and trudged upward for an hour and a half, first through a scrubby forest, then through the zone of bamboo. This area was established as a gorilla sanctuary in 1930, and as a forest reserve in 1939, contiguous with Albert Park. The forest on the lower slopes was cut from the reserve in 1950, raising the boundary from the 8,000 to the 9,000-foot contour line. This was an extremely unfortunate move, for the forest removed from protection was, as it turned out, the most suitable habitat for the gorillas. Attacked by a land-hungry population, a forest does not survive for long. Goats and cattle graze away the undergrowth, men fell the trees, and women turn up the soil. The gorillas must retreat ever farther up the mountain.

Walter's mountain camp is in the saddle between Mts. Muhavura and Gahinga in a little clearing near the upper limit of the bamboo at an altitude of nearly ten thousand feet. There are four huts, one of metal and three of bamboo, overlooking the plains of Kisoro and on exceptionally clear days one can see the Mountains of the Moon north of Lake Edward.

During the following days we were educated in the ways of tracking gorillas. We went up and down canyons, through the bamboo and dense brush, always seeking the spoor that a gorilla group leaves behind: a heel print in soft earth, like that of a man but wider, a vine stripped of leaves, a small hole scooped in the soil where the ape had dug for the tasty taproot of a borage. The guides followed the spoor closely, and, when it disappeared, Reuben waved the others to circle until they cut across the trail again. There was little talk, everyone keeping in contact by low whistles, like pensive birds. The guides scurried about, hunched over, darting glances here and there and back at us. They resembled bird dogs in search of their quarry.

It rained frequently, for the wet season had arrived. Here near the equator the year is divided into two dry and two wet seasons. January and February tend to be fairly dry and March through May wet; June through August are very dry, and from September through December the rains fall steadily. The slopes became slippery, and clouds settled in the saddle. We had found a small group of gorillas — one silverbacked male, one black-backed male, three females, and two infants. Once a downpour soaked us, and, as we continued on the trail of the animals, we suddenly came upon them resting snug and dry under the overhang of a rock. They immediately bolted up the slope on seeing us, but the large male remained behind to face us hidden behind a screen of bamboo. He roared. Reuben urged us to go closer, but I was content to view the shadowy figure from a distance of eighty feet. He roared again, then silently melted up the slope to join his group.

We could learn little about the undisturbed social life of the gorillas by such encounters. Doc and I soon decided that a gorilla study of the type we desired was impossible to carry out in this area. The dense undergrowth defeated us, just as it had done for Osborn and Donisthorpe before us and would do for Kawai and Mizuhara, two Japanese investigators, after us. All the same, the terrestrial gorilla, more than any other monkey or ape, is satisfying to study, for even if the animals themselves remain hidden their spoor can reveal much of interest. Doc tracked one group for seven continuous days during which the gorillas remained in an area half a mile wide and one mile long. The group meandered up into the bamboo and down into the forest, back and forth across its own trail, traveling about a mile a day. We found, by counting food remnants along the trail, that nearly half of their forage at this time of the year consisted of bamboo shoots, which the animals shredded or peeled like a banana, discarding the tough hairy outer bark and eating the tender white pith. They also favored the roots of a borage, the leaves of certain vines, and wild celery (*Peucedanum linderi*). By picking apart pieces of gorilla dung, which is three-lobed and resembles horse manure in consistency, we found the seeds of several fruits. In all, gorillas ate parts of twenty-nine different plants in the area. We also mapped nest sites and counted the nests, noting that of

106 nests a little less than half were on the ground, constructed of bamboo, vines, and brush. The others were up in the branches of trees and in the canopy of the bamboo, the latter delightfully springy and soft, better than any mattress and to me a great temptation to abandon the trail for a nap gently swaying in the sunshine.

Reuben and the other guides told us many things about the family life of the gorillas, but like the other Africans we met, they were not reliable. When it came to confirming the presence or absence of the apes or of pointing out the various food plants, our guides were invaluable, but when they were describing the behavior of the gorillas they simply assumed that it resembled their own. Thus gorilla males were said to bring food to their families, and a female about to give birth is supposed to leave the group and seclude herself.

While Doc followed one group, I searched for others. Kay and I climbed to the summit of Mt. Gahinga and looked down into the green crater with giant senecios growing on the slopes. Tattered clouds swirled in the depth of the cauldron, and a lone buffalo rested on a knoll massively chewing its cud. To our surprise we also found the dung of elephants at this high altitude. I climbed to the summit of Mt. Muhavura three times in three days, for I had seen fresh sign of gorillas at an altitude of 13,500 feet and hoped to encounter the apes on the open slopes. A small pond, perhaps seventy feet in diameter, marks the summit of the mountain, first reached by two German officers, Bethe and Pfeiffer, in 1898. I felt good on reaching these heights. The muscles of my legs were hard and corded with the effort of the climb, but I had lost the weakness that overcame me on Mt. Karisimbi. The views from the summit were immensely wide. Everything was great and free, and the air was alive over the valley, shimmering like a flame. About fifteen miles to the north, across the barren hills, was another forest in which gorillas lived. As in the Virunga Volcanoes, these animals too were isolated from contact with others by the cultivation that surrounded their home.

We had heard of the gorillas in the Impenetrable Forest — a name romantic and mysterious that inspired the imagination — but no one had studied them there. In 1912 a certain Critchley-Salmonson is said to have heard gorillas in this forest, but it remained for Pitman, a former game warden of Uganda, to bring out the first reliable information in 1934 and again in 1937. The animal behaviorist Niels Bolwig made a passing visit to the forest in 1959, and now Doc and I planned to survey this area for eleven days. Fangoudis, now a white hunter but formerly a miner, had spent five years in this forest searching for wolfram and other metals, and when we visited him, he pointed out to us where gorillas might be found and he also told us of the Batwa who know the forest as no one else.

We met our first Batwa at the Travellers Rest Hotel. He came up to us, clapped his hands, stomped his feet in a rather shuffling jig, and shouted "*eh haw, sawa sawa sawa*" with a hoarse voice. When we gave him a shilling, his broad wrinkled face broke into a grin and he dropped the coin into a little woven bark bag which he always carried under one shoulder. He was a beggar wrapped in a dirty striped blanket, but he had a certain dignity. He was also an excellent mimic, truly funny, for he played his roles with comic exaggeration, taking pleasure in miming those aspects of life which the Bantus around him took seriously. He was barely five feet tall, a full-grown gnome who belonged not here but in a forest glade.

The Batwa are just one of bewildering series of peoples living in this part of Africa. The tall Watutsi, famous for their spectacular dances, are found primarily in Rwanda and Burundi. Aggressive and warlike, they swept down from northern Uganda in the seventeenth century and subjugated the tribes around them, intimidating them by their very height. They are arrogant, beautiful people with features like the ancient pharaohs, but centuries of easy life have weakened their spirit, and to me they seem somewhat pathetic. Clothed in a white sheet and carrying a long staff, they watch their herds of cattle. Cattle are their wealth, their status, their love, their life. The cattle tend to be scrawny, ribs outlined sharply beneath their brown or white-splotched hides. They are bred neither for meat, for no cow is killed for food, nor for milk, but for the size of the horns, which in some sweep outward and upward for five or more feet. The power of the pastoral Watutsi collapsed when the winds of independence swept across Africa. The Bahutu, comprising 85 per cent of the Rwanda and Burundi population, turned on their former masters, and from 1959 onward the country was torn by tribal wars. The Bahutu are agriculturalists. No one knows how long they have occupied this area. Like the Bashi, Bahavu, Bahunde, and other tribes, to whom they are linguistically and racially related, they have been here for many hundreds of years. A thousand years ago Bantu agriculturalists were already growing millet in the hills near Bukavu.

The Batwa are pygmoid forest dwellers who live by hunting. They are somewhat taller and stockier than the pygmies of the Ituri Forest, but even the males average only about five feet in height. Once the Batwa were widespread in this area, but when the agriculturalists cleared away the forests, they withdrew westward into the mountains of the rift. A few cling to their ancient homes — the Virunga Volcanoes, the Kayonza Forest — but others have given up the life of hunting and roaming to become beggars and artisans, making clay pipes and bowls for trade. They speak the same language as the local Bantu, except for a few dialectical differences. No one knows the true history of their race. According to the Batwa, Woto, the fourth chief of the Bushongo, left his people and retired into the forest. He found himself very lonely and uttered an incantation. Thereupon the trees opened and sent forth a multitude of little beings. When asked what people they were, they answered: "Binu Batwa (We are men)."

High on the slopes of the rift, north and west of Bukavu, Doc and I visited a Batwa village with Christiaensen, a former coffee planter who was employed by a local research station to buy any

interesting animals the Batwa have killed. The village huddled
in a clearing on a steep slope surrounded by forest. Below
and far away the hills flatten out to disappear in the misty expanse of the Congo basin. Eleven huts, only six feet high, stood
in two rows like a series of beaver lodges. They were simple structures, transitory, of bent saplings tied to a central pole and covered with layers of grass. Inside, along the wall farthest from the
door the family slept on a bed of grass. A fireplace with a few
black and cracked rocks was near the center pole. Various pots
and baskets filled with beans, manioc, and other produce were
piled near the entrance. Outside, on a horizontal pole supported
between two huts, hung two dappled bush buck hides; two others
and that of a duiker were pegged out to dry on the hard-packed
soil. Several men had left to scour the forests for game, which,
when caught, would be eaten by the villagers or traded to the
Bantu for agricultural products. Three men cut apart the hind
quarter of a forest pig. They squatted by the red meat on its mat
of green ferns, while a woman kindled a fire. There were seven
women in the village; tiny creatures, in looks not full grown yet
complete. They wore sarongs of colored cloth and from their
wrists and ankles dangled copper bracelets. Each had a blue line
tattooed on the forehead, running down to the tip of the nose;
various notches and crosses were burned or engraved into their
cheeks, upper arms, and breasts. One woman had a broad color-
less tattoo burned like a necklace around her neck. The Batwa are
nomads, rarely living longer than one year in a place. When the
game grows scarce and shy, they move on. The undergrowth
which has stood in wait at the edge of the village retakes the
ground, the houses rot, and the history of the people returns to
the soil.

With the help of the African district chief, we found some
porters to carry our loads into the Impenetrable Forest; with us
also were four Batwa and a Bantu boy. Ephraim Busingye was
about twelve years old, a bright-eyed pleasant lad who had
learned English in a mission school and who was to act as our
interpreter with the Batwa. Our wives were not to accompany
us, for we were uncertain of the difficulties that lay ahead of us

in the forest. Jinny had gone to Kampala and Kay to the Lake Edward Hotel in Katwe. From the road our trail led for an hour into the interior of the forest. The path was steep and slippery, and all around us the large trunks loomed upward to vanish into the green world of the canopy. Finally we reached Batana, a knoll on which once long ago the trees had been felled. The Bantu chopped away the underbrush with their pangas, vicious-looking weapons with a blade in the shape of a question mark. We erected our tent in the clearing, and two of the Bantu, besides Busingye, built themselves a hut of saplings and ferns somewhat downhill from us. The Batwa constructed their shelter behind our tent, all the while chanting *eeee-ye-ye-eeeeeeeeeye* in exuberant fashion. A miner named Jack Collins had once lived on this knoll, but his hopes had crumbled with the price of the metals, and now only the rock foundation of his home remained, densely over-grown with creepers, the retreat of lizards and mice. But he had planted some rose bushes, and these survived, their red blossoms striking in this alien land.

It was my first visit to a rain forest. That evening I looked out over the miles of billowing foliage, black and motionless, symbol of all that is wild and untamed. The air was alive with sounds: the buzzing of cicadas, the faint twittering of bats, the high chirp-ing of tree frogs. The Africans talked quietly. Every little while one reached forward and placed a twig on the fire, and the flames reared upward casting a faint glow over our camp.

At night from the cozy warmth of my sleeping bag I heard the Batwa talking intermittently. They slept a few hours, then stirred to share a joke, laugh, only to slumber again. Once a Batwa played his *likembe*, a flat hollow instrument which pro-duces a simple series of sounds when the parallel iron bars on it are vibrated with the thumbs. It was a remote, haunting tune that seemed to come from another sphere.

Early in the morning Doc and I crawled off our camp cots and cooked our breakfast of oatmeal over the open fire. We were at an altitude of nearly seven thousand feet, and the air was chilly. Clouds had settled in the deep valleys, and the summits of the dark hills floated in a white sea. We sent two Batwa in one direction to search for gorillas. We took the other Batwa and Bisingye southward into the depth of the forest. The Impenetrable Forest is a reserve, established in 1932. Timber felling began in 1940, but the activity is largely confined to the peripheral and accessible portions. Only 100,000 cubic feet of timber may be removed each year by the ancient method of pit-sawing the huge boles into planks. The interior of the forest remains largely untouched by man, a primeval island surrounded by heavy cultivation. Only the Batwa call this forest their home, and they are as much a part of it as the pigs and the monkeys they hunt.

We hoped that this remnant of mountain rain forest would always remain a reserve, for if it were opened to agriculturalists the hills would soon be as denuded as the surrounding country is now. It is hard to believe that mighty forests once capped the mountains around the reserve. As recently as 1910 elephants and buffalo roamed where today not a tree remains. In 1860 the Bakiga tribe immigrated from Rwanda, and this, together with the natural increase of the population, raised the number of people in the Kigezi district from 100,000 to 600,000 in fifty years. The natives felled the forests and planted their crops, and when the soil was exhausted the land lay fallow. But each time a new growth of trees began to spring up the hills were burned over. Soon the plant succession was deflected to grasslands. Today the hills stand brown and bare under the blazing sun. Sometimes an African wanders by and sets fire to the rustling grass, either to encourage a new growth of green grass for his goats or simply from a pyromanical impulse. Then the air dances with the conflagration, and layers of smoke roll over the land. And at night lines of fire blaze disembodied in the sky.

The gorilla, the elephant and other forms of wild life cannot survive for long without the preservation of the vegetation. In Uganda the adaptable elephant is a good case in point. The elephant can subsist on the hot grasslands and in the cold mountain forests. But if its habitat is converted into fields and if it has to compete intensively with livestock, the animal can do nothing but retreat. In 1929 elephants occupied 70 per cent of all the land in Uganda; in 1959 their range had shrunk to 17 per cent, nearly all confined to forest reserves, national parks, and "sleeping sickness zones."

We followed the crest of a ridge in single file. Bichumu, the chief of the Batwa, trotted ahead. His kinky hair was white and his short beard scraggly. The skin over his knees was wrinkled like that of an elephant, and his old, frail body was wrapped in a torn khaki jacket. Like all Batwa, he carried two spears in one hand and a machete in the other. Behind him followed Bataka, whose sad, bulging eyes hid a bubbling spirit which found release in exuberant jumps.

Our progress along the ridge was rapid. Here the forest certainly was not impenetrable. The infinite line of lofty trees roofed the whole land, excluding the direct sun. Only some saplings survived in the gloom, with the result that the forest near ground level was relatively open. Here and there was a well of sunshine where some forest monarch had crashed to the ground in a tangle of splintered branches. Our footsteps fell muted on the carpet of leaves, and the steaming air was heavy with the odor of decay. An earthworm, over a foot long and with the diameter of a man's finger, disappeared in a rotten stump. A black line of army ants moved across the ridge, flanked by soldier ants ready to fasten their pincers into any intruder. The morning coolness had given way to midday heat. We rested at the end of the ridge and looked a thousand feet down into the valley which we had to cross. The forest at noon was still, and nothing moved. The rain clouds were banking up around us, as they tended to do every day. The forest seemed oppressive, with an overpowering heaviness.

The Batwa showed us many things. They pointed out the heavy trunk of a tree, called *ornukungashebeya*, used for making dugout canoes, and they showed us a vine, *olushuri*, from which they nimbly stripped the bark to weave little bags. But when we peppered them with questions "Do gorillas eat this plant?" "Do they nest in trees or on the ground?" our inquiries were met with stubborn silence or evasive replies. Bichumu, obviously distressed, finally told us: "If you call the animal you are seeking by name, you will never find it."

At several places along the trail we saw ingenious snares set by the Batwa for forest pigs. A sapling is bent over, and the noose of woven bark is concealed beneath a layer of leaves that also covers a little hole. The noose is held in place by a peg clamped into a split horizontal stick. When a pig steps into the hole, the split stick is depressed, thereby releasing the peg and thus the noose. The sapling jerks up and the noose holds the leg of the animal against a branch. The same trigger mechanism is also used in deadfalls, except that a heavy load of logs falls on the unsuspecting animal, pinning it to the ground.

Slipping and sliding, we descended into a valley, into another world. Few tall trees grew on the lower slopes and in the valleys. A riotous mass of vines, herbs, and shrubs had taken over. The Batwa cut a path with their machetes, but at times we simply crawled several feet above ground over the swaying mass of vegetation. Thorns tore at us, and vines tugged at our feet. There were stands of tree ferns with black stems and fronds that flowed gracefully outward from the apex. The light in these groves of ferns was translucent green, subterranean in quality. Spines covered the stems of the ferns and often, when we stumbled, our hands automatically reached for a stem for support. Roaches scurried about. The atmosphere was primeval, paleozoic, as when the world was ruled by tree ferns and amphibians. Small streams sought their way down most of the valleys, and from these we drank deeply and cooled our hot faces.

We continued on, up and down, the Batwa trotting ahead, Doc and I following as best as we could. The Batwa had a kind of rapport with nature: they seemed a part of their surroundings. The gurgling of water, the faint rustling of leaves, the swift flight of a flock of grey parrots, had ceased to be received by their senses, for they formed the heart of the senses themselves. Occasionally they stopped to examine the track of a pig barely visible in the decaying leaves, or they jabbed the end of the spear into the soil and then, digging with their hands, jubilantly extracted a wild tuber which they stuffed into their bags. Once the underbrush rustled, and both were transformed to stone, standing motionless, spears raised, ready to thrust. But the animal moved

Gorillas occasionally visit the zone of giant senecios *at an altitude of 12,800 feet on the slopes of Mt. Muhavura.*

A Batwa village in the mountains west of Lake Kivu.

ur guide in the Impenetrable Forest was Bichumu, the Batwa chief.

The Virunga Volcanoes, looking east from an altitude of 12,000 feet on the slopes of Mt. Mikeno. The jagged peak is Mt. Sabinio; the one with its summit in the clouds is Mt. Muhavura; the flat-topped peak is Mt. Visoke. Low tree heath and white-flowered Helichrysum *grow in the foreground.*

off without showing itself. At one time, they told us, several Bantu had gone out to spear pig. When the vegetation swayed they threw their spears, thinking that it was a pig. But it was a lone adult male gorilla. In the ensuing melee one Bantu was bitten in the buttocks. Delightedly the Batwa pantomimed a person trying to sit, only to jump up in agony holding his rump.

The Batwa enjoyed teasing the Bantu, in retaliation, I suspect, for being held inferior by them. In our camp the Bantu refused to sleep in the same hut or share food with the Batwa, and they ordered them around whenever possible. Yet the Batwa were gay with an uncomplicated cheerfulness that contrasted sharply with the rather sullen demeanor of the average Bantu in our employ. Is it possible that the hard but unfettered life of the hunter brings with it a freedom of the spirit that the sedentary agriculturalists have lost?

One Bantu had told me that gorillas grab flying spears when hunted and throw them back at the attacker. When I asked Bishumu about this, he smiled and replied: "This is a fable. We tell such tales to the Hutu and they believe them. The gorillas fear man. They bark and roar and run away."

There is little wild life to be seen in a mountain rain forest. Many creatures are nocturnal and shy. Although I heard pigs rustling away at my approach, I never saw any. Elephants are seen in the western part of the forest but for some reason failed to penetrate as far as our camp. Once we encountered a band of about ten Hoest's monkeys feeding in the undergrowth. These semi-terrestrial monkeys are strikingly colored, with a black body, a rusty brown back, and a white bib covering the chin and sides of the face. Like large cats, they scampered away across the swaying mass of vines. The other species of monkeys in the area seemed to prefer the forest's edge rather than the interior. There was the blue monkey, which we again saw later on Mt. Tshiaberimu, and the red-tailed monkey, a pert little creature with a white spot on its nose and an almost iridescent red tail. The most flamboyant of the monkeys was the black and white colobus. Small groups, consisting of from three to fifteen animals each, sat in the crowns of the trees and fed on buds, leaves, and blossoms.

When I approached them, the males sometimes turned their black faces towards me and opened and closed their mouths, producing a smacking sound. Then they fled, their mantle of white flapping in the air and the long tail with its enormous tuft of white hair flowing behind.

Even the birds were inconspicuous. Many remained in the forest canopy, fleeting black shadows against the brilliant light. Occasionally a black and white casqued hornbill flew with whooshing wings from one tree to another, emitting its loud, nasal *kwo-kwo* call. I searched for the nest of the hornbill but failed to find one. A female usually lays her eggs in a hole high up in the trunk of a tree. The male carries pellets of earth mixed with his saliva to the hole, and the female builds herself a prison, plastering the opening up from the inside until nothing but a narrow slit remains. For about four months she stays in this hole, molting, incubating her eggs, and raising her young until they are ready to fly. All the while the male feeds her on fruits which he faithfully brings to the opening and regurgitates for her.

Once I saw a small snakelike creature writhe on the ground at my feet. It reminded me of when as a boy I had attempted to catch lizards and only succeeded in holding the detached tail wriggling furiously in the palm of my hand while the animal itself escaped. Much to my surprise, what I caught here was not a snake or lizard but an earthworm brilliant blue in color. It was the most athletic and colorful earthworm I have ever seen; but as so often happens when one tries to take a beautiful thing away from its natural surroundings, the alcohol in which I attempted to preserve the worm dissolved the color, and my jar held nothing but a flaccid, pinkish piece of flesh.

Gorilla signs were widely scattered and far less abundant than at Kabara. There were small piles of bark from the fronds of tree ferns, which the animals had detached to get at the tender pith, and bare vines from which the bark and leaves had been stripped. Unlike the region of the Virunga Volcanoes, there was no bamboo or wild celery in this part of the forest, and the animals subsisted on a different diet of plants. In all we collected twenty-seven plants eaten by gorillas, and of these nine were also eaten in the Virunga Volcanoes. Most of the nests which we located were old, with only the broken dead branches to show where the apes had rested. A nest may remain discernible for as long as a year. Some nests were on the ground, others several feet up in the bent crowns of shrubs and saplings. A few gorillas had constructed their nests in the branches of trees at heights of forty-five feet where they perched like hawks on their eyries. I counted 179 nests in the Impenetrable Forest, and of these about half were in trees.

We found no gorillas the first day or the second, but on the third we heard a scream across a valley. Our approach was silent, and then I saw the animals about fifty yards away on the opposite slope. The Batwa fidgeted and would not remain quiet, and finally I sent them home. The gorillas had not yet seen us. A splendid silverbacked male lay on his back among the vines. Indolently he gathered a handful of vegetation and plucked several leaves with his thumb and index finger. Then he pushed the leaves into his mouth, chewing lazily, the sun sparkling on his bare chest. Three females and a juvenile foraged silently, now visible to me, now disappearing in the undergrowth with only the swaying of branches to mark their progress. A female climbed a tree to a height of twenty feet and sat in a crotch. She pulled a vine close to her mouth and bit off the leaves. Then she slid down the trunk hand over hand and went to sleep on her side with her head cradled on her arm. For about four hours I watched the gorillas resting and feeding, until they moved from my sight.

While I was quietly creeping up the slope, trying not to alarm the gorillas, there was a sudden snakelike hiss at my shoulder. I started. A black-billed turaco flew up, flashing its brilliant crimson wings. Then this crow-sized bird hopped along a branch, chattering like a squirrel. Once I had reached the crest of the ridge, travel was again easy. Every few minutes I stopped and took a bearing with my compass. Without a compass I would have been lost in this forest. The ridges and valleys are not aligned in any direction, and each gradually and unnoticeably veers to one side or the other. Only the Batwa invariably seemed to know the way. But I remember once when Busingye, two Batwa, and

I traveled far to the southwest to a part of the forest with which the Batwa were relatively unfamiliar. We came to a fork in the trail. The Batwa insisted that camp was in one direction, my compass pointed in the other. I was tempted to disregard the compass and follow their urging, and then decided against it. Busingye trotted at my heels pleading: "Master, you are leading us astray." Much to my relief, I was not.

On one occasion there was a joyful shout from a Batwa, and the others rushed over. He had discovered bees hovering by a small hole in a tree about twelve feet above ground. Immediately the Batwa wedged a sapling against the trunk and climbed up. They lit a dry, rotten piece of wood and blew the heavy smoke into the opening. They made little progress pecking away at the bark with their spears in an attempt to enlarge the hole and decided to return the following day. During the next attempt their spears were fitted with a chisel-like point. The chips flew as they chopped, and they were coughing and spitting in the dense smoke. One Batwa reached into the tree and hauled out the combs dripping with golden honey. They ate and ate, laughing and swatting at the bees that buzzed angrily around them. Some combs held white grubs, and these were eaten too. Only the wax was spat out. The Batwa had honey all over their arms and chest, and the bees landed on their skin and stung them. Then they licked their fingers and arms and stuffed the extra honey into the small calabashes they carried.

The rainy season was still with us. A sudden wind squall that lashed the canopy high above us was frequently the harbinger of

a storm. Then there was silence, a deadness in the air, as if the forest stood implacably waiting. Even the shrilling of the cicadas ceased. Sometimes a bird called far away, a brief, lonely sound that intensified the silence by its very contrast. Soon a mysterious roaring, at first faint and elusive, then loud and insistent, approached closer and closer until the rain wrapped us in its grey folds. We stood huddled against the bole of a tree, peering out from beneath the plastic tarp we held over our heads. The rain plunged down in a solid sheet, and the trees across the valley were blurred. These violent storms ceased as suddenly as they began. The Batwa shivered in the cool air and built a fire. Bataka emptied the contents of his bag on the ground. In it were a dry corn cob, a small home-made knife, a matchbox wrapped in a piece of cloth, some tobacco leaves, a wooden pipe, and two sticks. He took the two small sticks, each about 2 feet long and carefully wrapped in leaves and tied with vine. The point of one stick he rotated rapidly on the other by rolling the stick back and forth between his palms. The friction produced fluff-like shavings. Faster and faster Bataka worked; his eyes bulged, and beads of sweat sprang from his forehead. After a few minutes the shavings smoked and began to glow. He tore a piece of cloth from his coat and together with some dried leaves held it to the glowing shavings. Now I knew why his coat consisted only of seams on which a few tattered patches of cloth remained: long ago the coat had ceased to serve a protective function and had become the mere suggestion of a garment. Soon a cheerful little fire burned. While we were warming our hands over the flames, a gorilla screamed in the distance. We left the fire and crept down into the valley where the animals were feeding. Then we heard a curious gorilla call which I have not heard since — a call which seemed to me to be like the neighing of a horse. Other observers have reported similar sounds, but we have no knowledge of its meaning.

I ordered the Batwa to stay behind as I crept closer. Two gorillas disappeared in the undergrowth. Thirty feet ahead a bush shook, and I could barely distinguish the dark form of a gorilla feeding quietly. I circled the animal silently, one step at a time. But observations were impossible, and just as silently I crept away.

The Batwa were upset by my inquisitive approach to the gorillas, but whether this was for my sake or because they were afraid for themselves, I was not certain. When we followed a fresh trail, Bataka repeated over and over like a litany: "If you follow this trail, the gorillas will kill you."

Occasionally the Batwa deliberately lost the trail. In June, 1960, when I returned to the Impenetrable Forest with Dr. Niels Bolwig, then of Makerere College, I had learned to track gorillas almost as well as the Batwa. Now, when they veered off the spoor, I simply continued on alone, and soon they caught up with me, troubled but willing to follow.

Once I tracked a group of fifteen animals for four days. No rain had fallen for several weeks and the leaves were dry and crackling underfoot on the ridges. The damp valleys were unchanged. I rarely caught more than fleeting glimpses of the group in the dense undergrowth, although I tried to watch it daily. Yet I learned much from the trail. I noted, for example, that the animals seemed to avoid contact with water. Three times they crossed a small stream, barely eight feet wide and at most two feet deep. Once they marched across a fallen log, next they broke down three tree ferns so that the stems bridged the stream, and the third time several gorillas jumped onto a barely submerged rock in midstream before gaining the opposite bank with another leap. I also found, as we had noted earlier, that the gorillas fed primarily in the lush valleys, but moved to the upper slopes and ridges for sleeping.

The presence of chimpanzees in the Impenetrable Forest intrigued us greatly, for we found the nests of gorillas and chimpanzees on the same ridge. Unfortunately, we never had the luck to see these two closely related species together. I examined the nests of the chimpanzee and found that they were similar to those of the gorilla; only the nest location and the dung revealed the identity of the animal. No chimpanzee nested on the ground; their nests were usually between fifteen and eighty feet above the ground in the canopy of trees. The dung of chimpanzees is totally unlike that of the gorilla, in shape tending to resemble that of man.

Toward the end of our stay in the Impenetrable Forest, we assessed our effort. We had found fresh nest sites at five widely scattered places, and these probably represented five groups that varied in size from two to fourteen animals. The gorilla population seemed more sparse than in the Virunga Volcanoes. After visiting various parts of the forest, we estimated that about one hundred and fifty gorillas lived here. We had learned much of their nesting habits, of the food they ate, and of their general activity pattern. But we realized that, as at Kisoro, prolonged observations of groups would only be possible when the animals crossed the open valleys.

Before leaving the area, we wanted to check the eastern part of the forest briefly, and the quickest way to do this was by car. When it came time to start, the Batwa were afraid. Three of them finally climbed into the backseat, giggling nervously and clutching their spears. It was their first trip in a car. But Bishumu stood on the road, a defiant look on his face. The others leaned out the door and beckoned him, laughing and shouting with uneasy courage. Bishumu only shook his head, a proud, erect old man afraid of the alien civilization that so recently had encroached on his home. Suddenly he sat down in the road with his back to us. We had to leave him there. When I looked back, he sat motionless, a small brown figure alone on the yellow road, with the green forest all around him.

CHAPTER 3

Interlude in the Plains

I found a little knoll and sat there, binoculars in hand, in the morning sun. Mt. Tshiaberimu floated dimly on a bank of mist, and far out over the shimmering surface of Lake Edward hovered black columns of lake flies. Across from me, on the other side of the river, was a sand bank crowded with water birds. Cormorants stood there in dense ranks, some with wings spread and with yellow throat pouch vibrating rapidly. Occasionally two cormorants greeted each other with elaborate ceremony. Weaving their heads back and forth, they advanced with mincing steps towards each other and, with chests almost touching, stretched their necks skyward.

It has been suggested that man and other animals find it difficult to adapt to a biotope unlike the one in which they were raised. For example, pygmies are said to dislike the open plains, and Europeans feel uneasy and hemmed in among the towering trees of the rain forest. I suspect that the statement has considerable truth; at least, I always felt a certain relief when, after spending several weeks in the forest, the views were wide again. After working in the Impenetrable Forest, we drove northward through the Rwindi game plains of Albert Park. Rwindi, near the southern end of Lake Edward, is the only place with regular tourist accommodations within the park. And it is the central location from which to see the herds of buffalo, elephant, hippopotamus, waterbuck, kob antelope, and such other game as lion, leopard, and warthog, on the grasslands and by the palm-fringed rivers.

We followed the road past Rwindi, out of the rift valley and northward through the mountains, and descended to the town of Beni. Beni lies at the edge of the Ituri Forest, a vast sea of green that stretches to the horizon. The town has several stores, operated by Greeks who do much of the trading in the eastern Congo, and the small Majestic Hotel at which we stayed. As the sun was setting and the plains were already in deep shadow, we saw the Mountains of the Moon. They rise nearly seventeen thousand feet from the valley floor, an upthrusted peneplane heavily eroded and with glaciers clinging to the bleak rock. We watched the glaciers flush a deep gold, then slowly fade as the earth leaned away from the sun. We had seen a rare sight, for often the Mountains of the Moon remain hidden in the clouds for weeks on end, giving no intimation to the casual traveler that they exist. The natives aptly call them *ruenjura*, "the place of rain." During the first century A.D., Diogenes, a Greek, supposedly went inland from Zanzibar for twenty-five days to the vicinity of two great lakes and a snowy range of mountains from which the Nile draws its source. The Syrian geographer Marinus of Tyre recorded this information, and in 150 A.D. Claudius Ptolemaeus, commonly called Ptolemy, drew his famous map showing the Nile flowing northward from the *Lunae montes*, the Mountains of the Moon. Yet the world had to wait seventeen hundred years before the account was verified. Although Henry Stanley is credited with discovering the mountains in 1888, other Europeans had seen them before him. In 1864 Sir Samuel Baker, the discoverer of Lake Albert, noted a range in the distance and called it the Blue Mountains; two members of Stanley's expedition, Parke and Mounteney-Jephson, spotted the glaciers from Lake Albert a month before Stanley himself did so.

No gorillas are found in the Mountains of the Moon, but chimpanzees roam along the lower slopes. We allowed ourselves two days early in May to explore the mountain massif. While Doc searched for chimpanzees near the base, I followed the Butagu Valley. I saw where the apes had eaten the pith from the stem of a wild banana, and I heard the animals hoot in the distance. Climbing rapidly with one guide, I passed through the bamboo zone, upward into stands of tree heather where the sphagnum moss was several feet deep, the trail a morass, and the light shone eerily through the silver-grey lichens that festooned the heather branches. We left our packs at the Mahangu hut and continued into the zone of giant senecios. At an altitude of some thirteen thousand feet we came to *Campi ya chupa*, the "Camp of the Bottle," so named because the German zoölogist Franz Stuhlman left a bottle on this knoll to mark the highest point reached by him in 1891. We continued on a little way, then headed back to the hut. The glaciers were obscured by clouds, and to my left in the shady depth of a valley nestled Lac Noir, its black surface calm as death. We returned to the lowlands the following morning. I had accomplished nothing; in fact, my trip had been foolish, yet I was pleased to have been on the mountain.

That evening we met Micha, who had come to inspect the northern portion of Albert Park. As we sat on the veranda of the Mitumba Hotel at the base of the mountains, drinking gin and tonic and talking of this and that, Micha asked: "Have you been to Ishango? No? But then you must go there. It is the most beautiful spot in Parc Albert."

We followed his advice and in the late afternoon on May 8 left the highway near the Congo-Uganda border. The road south to Ishango consisted of two ruts that led on and on across the undulating grasslands, with here and there a scattering of acacia trees. Far ahead a herd of fifty elephants moved through the golden stillness, majestic forms, rulers of a bygone age. To our right the rift escarpment rose abruptly from the valley floor, the slopes indistinct in the slanting rays of the setting sun. The highest point of the rift, some seven thousand feet above us, was Mt. Tshiaberimu, the northernmost extension of the mountain gorilla range. To the north of this mountain massif the altitude of the escarpment dropped abruptly. I eyed Mt. Tshiaberimu with interest, for it was the site of our next survey.

Ishango is surely one of the loveliest spots in Africa. It lies on the northern shore of Lake Edward where the Semliki River drains the fifty-mile-long lake and meanders northward to Lake Albert and the White Nile. Two rest houses stand near a bluff that overlooks the lake. We carried chairs to the rim of the bluff and watched the river some two hundred feet below us, curving gently toward the escarpment and flowing past brush-bordered banks and occasional euphorbia trees that stretch their dark green, succulent branches toward the sky. A small, tear-shaped sand bar near the mouth of the river was a favorite roost for a large number of water birds. About two hundred cormorants crowded together there; when they flew downriver at dusk, flocks of pelicans, both pink-backed and white, arrived in single file, one behind the other, with stately wingbeats. These grotesque birds settled on the island, the white pelicans on one side, the pink-backed ones on the other.

Below us in the reeds rested a hippopotamus, breathing heavily, a deep gash in its greyish-pink hide. Others lay submerged in the river, only their nostrils and bulbous eyes peering above the water. Occasionally one twitched its ears and emitted a deep *ho-ho-ho* that reverberated from river bank to river bank. Ten young hippopotami ventured onto a sand bar far downriver and like overgrown puppies chased each other around and around, their cavernous mouths wide open. A herd of eight tawny waterbuck filed from the brush and descended a narrow trail in single file. At the water's edge the buck looked around as if sensing danger, then lowered his long, curved horns and drank beside his does. Twenty elephants also came to the river to drink; we saw them as vague silhouettes in the rapidly fading light. After that there was nothing but the immense quiet of the night.

Although we left this enchanting spot early the following morning, I was able to return again in September, 1960, with Dr. Jacques Verschuren, then the biologist of Albert National Park. As I ambled alone along the shore of Lake Edward near the mouth of the river, I kept on the alert for hippopotami. The great bulk and spindly legs of the animals belie the speed with which they run on land. And when a hippopotamus feels itself cut off from its watery retreat, which it leaves to graze at night, it may attack. Frank Poppleton, the warden of Queen

Elizabeth National Park, adjoining Albert Park on the east, told me that of ten Africans who were killed in his park during the past few years, hippopotami accounted for five, buffalo gored four persons, primarily women gathering firewood in dense brush, and an elephant killed the last. Many hippopotamus trails worn into the hard clay by generations of ponderous feet led from the lake and the river onto the plains. Near these trails and usually by a bush were mounds of dung. The hippopotamus has the curious habit of urinating and defecating at the same time, and with its rapidly beating tail it scatters the faecal matter. Perhaps the dung and the urine serve as sign posts, indicating to others of the species who has been there; it resembles the way dogs stake out their property by leaving their mark on fences and trees. Count Cornet D'Elzius, who was then in charge of the Rwindi camp, told us a local fable to account for this habit:

God made the hippopotamus and told it to cut the grass for the other animals. But when the hippopotamus came to Africa and felt how hot it was, he asked God for permission to stay in the water during the day and to cut the grass only at night. God hesitated to give this permission, for the hippopotamus was apt to eat fish rather than cut grass. But the hippopotamus promised not to eat fish and was thus allowed to remain in the water. Now when the hippopotamus leaves a pile of dung, it scatters the pile with its tail, thereby showing God that there are no fish scales in it.

An adult hippopotamus, weighing as much as two tons, must cut a prodigious amount of grass to feed its vast bulk. Given protection from hunters, the number of hippopotami increased tremendously in Queen Elizabeth Park and Albert National Park with some 33,000 animals crowding the lakes and rivers. Even the sporadic anthrax epidemics from which the animal suffers fail to keep the population in check. In Elizabeth Park the problem became acute, with too many mouths sharing too few blades of grass. All land near the water was heavily overgrazed, and erosion followed. The animals had to go farther and farther afield to find a meal. Park authorities were then faced with a difficult problem, one that will ultimately plague all parks that harbor

herds of large game in small areas. They knew that their park was in danger of turning into a desert, yet they did not want to shoot the excess animals, for was not the purpose of a park the preservation of game in a natural, undisturbed state untouched by the hand of man? And yet man the hunter has been a part of the ecological scene for thousands of years, and most parks are much too small to be considered self-sustaining units. When their home becomes crowded, the animals have no place to go, for man hems them in on all sides. The Uganda authorities decided to shoot a certain number of hippopotami both in the park and around the periphery, thereby saving the land and providing meat for the Africans. The lesson from all this is clear: parks cannot be left to themselves, they must be properly managed.

I found a little knoll and sat there, binoculars in hand, in the morning sun. Mt. Tshiaberimu floated dimly on a bank of mist, and far out over the shimmering surface of Lake Edward hovered black columns of lake flies. Across from me, on the other side of the river, was a sand bank crowded with water birds. About eight hundred cormorants stood in dense ranks, some with wings spread and with yellow throat pouch vibrating rapidly. Occasionally two cormorants greeted each other with elaborate ceremony. Weaving their heads back and forth, they advanced with mincing steps toward each other and, with chests almost touching, stretched their necks skyward. Two sacred ibises, looking important in their pure white plumage and black-feathered head and neck, stalked back and forth. Nine wood ibises stood around with preoccupied mien, their bare crimson heads and yellow bills strikingly set off by their white and black plumage. When a pink-backed pelican dropped a few scraps of food, a wood ibis raced over and snatched up the morsel. Off to one side, on a patch of green grass, stood a mixed assortment of birds: two grey herons, one goliath heron, four white egrets, one yellow-billed heron, one saddle-billed stork, thirteen little egrets, and eight marabou storks. I have a special liking for the marabou storks with their bare reddish and black-splotched head and neck. They are far from beautiful, but they have a certain philosophical elegance about them as they pace gravely along. Their habits are those of a vulture rather than of a stork. During the heat of the day they soar at a great height, riding the thermals without effort. In the town of Katwe at the northern end of Lake Edward they are tame as dogs as they stand around the garbage heaps with bowed heads and hunched shoulders waiting for scraps of offal.

A varan lizard about three feet long swam toward the birds on the river bank. It crawled onto the sand, belly dragging on the ground, and waddled toward the cormorants, which simply stepped aside. A pelican sidled by and eyed its distant relative. For ten minutes the reptile wandered unmolested among the birds, then re-entered the water and swam away.

At my feet I found a piece of quartz that looked as if it had been chipped by man. Now that I had learned to see, I soon found a handful of broken arrowheads, scrapers, chips, and cores from which flakes had been struck. Long ago hunters had sat here and manufactured their weapons while watching the plains in the hope of surprising a waterbuck or some other creature as it used the narrow paths that led up and down the bluffs. Dr. Jean de Heinzelin, who had excavated an ancient village site at Ishango, told me that some of the stone tools may have been made as recently as one hundred and fifty years ago. But primitive man had used the site long before that. Some nine thousand years ago a village of mesolithic hunters and fishermen was located here. These men used spears to kill their game and harpoons to slaughter hippopotami; they also may have been cannibals. Suddenly the earth exploded around them, creating huge craters in the floor of the rift valley and on the slopes of the escarpment. About two hundred craters were formed over a period of years, some more than a mile in diameter and over one thousand feet deep. Today forest covers the rim of the craters and lakes lie in the bottom of some. They are one of the major attractions of Queen Elizabeth Park. At the end of the dry season, when the grass is again green, many elephants move into the craters and, viewed from the rim, resemble grey beetles moving over a green carpet. But to primitive man it must have been a time of terror as the earth shook and his huts were buried in grey ash. The village at Ishango was abandoned, and no new settlement grew up for several hundred years.

My days at Ishango and in other parks of East Africa gave me unforgettable glimpses of the world of nature as it was before the coming of modern man. To see African wild life in all its abundance and variety, living as it has always lived, was one of the most priceless experiences I have ever had. When I spotted a pride of lions resting lazily in the shade of a bush, when I watched warthogs trotting away, ugly yet curiously appealing with their self-important air, when I enjoyed the spectacle of hundreds of hissing vultures covering a cadaver and jackals and hyenas dashing into the seething mass to snap up some morsel, when I saw these and many other sights it was diffiicult to realize that most of the large game herds of Africa have been exterminated or drastically reduced and that very little wild life remains outside national parks and reserves.

Many persons still imagine Africa as a land teeming with vast herds of big game. Until World War II this was true in many parts, but today such aggregations exist only in the memories of the older generation. There is a little zoo in Entebbe, Uganda, where the game department houses orphaned animals before their shipment to zoos. As I saw the crowds of Africans delightedly looking at the creatures which they had never seen in the wild, it struck me how drastic the decline of game has been.

The causes behind the destruction of the animals are many. Illegal hunting by native poachers amounts to a large-scale commercial industry. In the past the natives hunted only for the pot. Now there is a sizable black market for dried meat, for elephant ivory, for rhinoceros horn, which has supposed aphrodisiacal qualities, for wildebeast tails, which make good fly whisks, and for giraffes, whose sinews are used for bow strings. In a one-hundred-square-mile area of Kenya, the ecologist Darling found that from one hundred to two hundred big animals were being poached each week. The poachers employ cruel and highly wasteful methods in catching their prey. A wire noose attached to a heavy log is often used for large animals like the elephant. In Queen Elizabeth Park we saw an elephant which the warden had to shoot because a wire cable on its leg had bitten through to the bone, leaving a festering wound. Pit traps are dug along trails, and hunters with poisoned arrows hide by water holes to take their toll as the game comes to drink. Sometimes traps are not inspected for many days, and the animal dies in agony, to be devoured by scavengers. Occasionally whole herds are driven to lines of snares that stretch over a mile of terrain. After a successful drive the poacher may not be able to dispose of all the meat, and many of the creatures are hamstrung, leaving them alive but unable to walk, to be collected later or not at all.

It has been jokingly proposed that a monument be erected to the tsetse fly, for this little insect has saved Africa's wild life by preventing agricultural and pastoral expansion into many areas. The tsetse fly feeds on the blood of wild game, which is a carrier of a deadly protozoan. Wild animals are resistant to it, but in cattle it causes a fatal disease called nagana. Other species of the fly carry sleeping sickness to man, but this disease has now been largely controled. Governments have long tried to eradicate the tsetse fly in order to open up the lands to human occupation. One approach has been to shoot all game animals with the hope of depriving the fly of its food supply and eliminating the host for the protozoan. In a recent two-and-one-half year period, 28,000 water buck, bushbuck, duiker, and other animals were slaughtered in one district of Uganda in the name of tsetse control. The meat was wasted. Over half a million wild animals were killed in southern Rhodesia in a twenty-five-year period for the same reason. Between 1927 and 1958 some 31,966 elephants were shot in Uganda in various control measures, a period in which licensed hunters accounted for only 8,170 elephants. Recent studies have shown that various small species of animals may act as vectors for the fly, making game slaughter at best a dubious method of tsetse control.

Whenever man occupies the land, the game decreases, for the animals are hunted and their habitat is destroyed. As Sir Julian Huxley has pointed out, the plains, savannahs, open woodlands, and other habitats in Africa are ecologically so brittle that intensive cultivation or heavy pastoralism rapidly turns them into a desert, soon without value to either man or beast. The African population is expanding, pushing even farther into the remaining wild life areas. What can be done to save a remnant?

The preservation of wild life ultimately depends, of course, on the African himself. The agriculturalist cannot be asked to save the game, for he needs the land and he craves the meat, and for the sake of his own survival he cannot tolerate elephants and buffalo in his fields. The pastoralist prizes number of cattle and goats above all else, and as his herds increase the wild game must be eliminated to conserve the amount of forage available to them. The plight of Africa's game has aroused international attention. Experts in wild life management like Grzimek from Germany, Darling and Huxley from England, and Buechner, Petrides, Talbot, Longhurst, and others from America have visited Africa to find a solution to the problem. In Nairobi the Kenya Wildlife Society was formed, and the New York Zoological Society started a wild life fund. A few years ago the future of the game looked very dim, but a belated, intensive effort by scientists has produced some tentative answers which, I believe, may eventually insure the preservation of some of Africa's unique fauna.

As a guiding proposition, it is realized that at present the African will preserve the game only if he derives direct material benefit from it, and that wild life is a major resource which requires proper utilization. Recently an important point has been demonstrated: many areas can produce a greater crop of wild animals than of domestic ones. For example, the land of one ranch in Southern Rhodesia yielded four pounds of game per acre as against three pounds of cattle. In Uganda it has been found that the buffalo will thrive and gain weight on natural grasses which will not maintain the weight of cattle. Game-cropping, or "harvesting" the annual surplus of wild animals for food on a sustained yield basis, in partnership with the local Africans, seems to be the best solution for maintaining herds of game outside of national parks. Such a scheme, with all of the proceeds going to the Africans, provides meat and hides while preserving the habitat. Tribes near game management areas also benefit from tourism, which is already the second largest source of income in Kenya. Hunting and photographing safaris bring in added revenues.

Ecologists have also discovered that in some areas a certain number of wild animals actually improves grazing for cattle. The diverse wild species of hoofed animals have achieved a wonderful balance within their natural community. Some graze on one type of grass, others graze on another kind; some browse on low bushes, others on high ones. Cattle, on the other hand, graze only on certain grasses. The plants which are not eaten by cattle spread at the expense of those which are.

One urgent need is to teach the Africans the value of wild life. We in America can speak from experience of the fate of game which is not properly conserved. The vast flocks of passenger pigeons, the Labrador duck, the heath hen, and others are gone, and of the thundering herds of bison, caribou, and bighorn sheep only remnants remain. According to Huxley, "conservation must become a central feature of policy. The emergent African nations must come to learn the harsh lesson that without proper conservation of soil and water and natural vegetation their lands will become unprofitable and useless, and also the helpful lesson of the positive values of their natural resources, including wild life and natural beauty."

The whole problem is one of human ecology. Man is conquering the diseases that once kept his population in check, and he is spreading his sway, exterminating other animals and exhausting the soil. With the same mentality that once enabled him to vanquish the lion and the bear, he is trying to subdue nature, sacrificing the eternal for the expedient. The destruction of the earth lies at his whim and cunning, yet he does not realize, does not feel, that he is not separate from but one with plants and animals, rock and water. He is as dependent on them as the protozoan, the tsetse, and the gorilla. By setting himself apart from the ecological community man has become a tyrant of the earth, but a tyrant who surely will fall if he succeeds in winning the struggle for existence.

CHAPTER 4

Mt. Tshiaberimu

I had not yet visited the northern part of the Mt. Tshiaberimu massif, and to do so I would have to spend a night on the trail. I hiked northward through the fields and villages in order to save time before entering the forest on my journey back to camp. In a country where officials rarely leave their offices and still more infrequently step off the roads, I, as a lone white man with a pack on my back, presented an enigma to the villagers. They crowded into the doorways of their huts and watched me silently, for they were distrustful of all strangers. Children followed me giggling and laughing, daring each other to go closer. When I suddenly stopped and looked back at them, they screeched and fled, their slender brown legs whirling like windmills. Soon this became a game between us, and the ever-increasing horde of children ran at my heels over the hills until I disappeared into the bamboo.

A sojourn to the highlands west of Lake Kivu and Lake Edward gives one the feeling of living up in the air. The climate is temperate, and even when the midday sun is straight overhead, one breathes easily. Many Europeans have settled in the rift mountains. There they grow tea and white-flowered pyrethrum, used in making insect powder. On green grass meadows graze Holstein, Frisian, and other dairy cattle where only a few years ago forest stood and gorillas roamed. The soil in the uplands is deep and fertile, making permanent cultivation possible. Much of the original vegetation has already been destroyed, leaving only a few remaining tracts high on the hills and in forest reserves. Still the population, both European and African, is pushing its fields ever upward. One day while Doc and I were searching for gorillas in the hills west of Bukavu, we drove to the end of a dirt track miles from any human habitation. All around us was bamboo, and we were shivering at nine thousand feet. Suddenly a field opened up before us and on it stood a little board shack. The day was but a thin solution of the night as the icy winds whipped stray clouds low over the ground. A Flemish couple with three small daughters was homesteading this lonely piece of land even though the Belgian government did little to encourage such permanent settling. They served us coffee which we drank while standing by the stove. They had tried to raise pigs and failed, and their luck now rode on a crop of turnips.

There were many missions in these mountains, imposing structures of brick covering the tops of hills with native villages hovering in attendance on the flanks. Catholic, Baptist, and Seventh Day Adventist followed each other with bewildering confusion, many living in a state of mutual hostility. The prestige of the Christian church in Africa has greatly declined because of the intolerance that the various sects show toward one another. Catholic missions in the Congo have been highly favored by the government, and from 1925 to 1945 they had the monopoly on subsidized education. I had little sympathy for these missions, but I admired their efforts to school the Africans, and on several occasions we appreciated their hospitality.

In mid-May Jinny went to Rhodesia to visit some friends, leaving Doc, Kay, and me to study the gorillas on and around Mt. Tshiaberimu, a peak which rises ten thousand feet on the rift escarpment. Although the boundary of Albert Park dips briefly away from the shores of Lake Edward to bring this mountain under its protection, the area is rarely visited by park officials. In recent years only Dr. Verschuren had spent a few days there. We approached Mt. Tshiaberimu from the west on May 16, walking in from the village of Masereka. Our porters marched ahead, cheerfully shouting to the women working in the fields. It was a pleasant land, fresh and green and covered with beans, potatoes, peas, and wheat. The small plots circled the steeply undulating hills along contour lines, each terrace separated from the one below by a hedge of elephant grass to prevent erosion. The villages huddled on the ridges, each round hut crowding its neighbor as if afraid of falling off. All forest had long since disappeared, and only coppices of Australian wattle grew in sheltered ravines, brown harriers circling lazily above them. Contour planting was strictly enforced by the Belgians, and it was also through their edicts that the scattered huts were concentrated into villages and that outhouses were located below the level of the main hut.

Mt. Tshiaberimu loomed ahead of us beneath black clouds, heralding the type of weather we were to have for the coming two weeks. After four hours of hiking we entered the bamboo forest. The rain began. The ground was soft underfoot, carpeted with leaves, and unbelievably slippery. We balanced on rocks across the Kibio River, which disappeared downstream into a dark tunnel of slanting bamboo stems on its long journey to the Congo River and the sea. We labored up Mt. Tshiaberimu and at last came to a small clearing in the bamboo. Dr. Verschuren had once camped here, and we decided to do so too. We dug a small platform, and the porters cut some poles which they set upright into the earth and tied together with pliable roots and vines to make the crude walls of a hut. Over the top we stretched some tarps; wooden crates served as shelves for our food, and low benches of split bamboo provided seats. Doc erected his one-man tent, and Kay and I each had a jungle hammock which we stretched between the sturdy *Podocarpus* trees that shaded our campsite.

Dr. Verschuren had lent us two of the Africans in his employ, because they knew this part of the country. Kieko, a quiet, friendly boy with a shy smile, was to chop wood, wash dishes, and do the other chores around camp. Christoph was to be a sort of major-domo, arranging for porters and acting as translator, for he could speak a few words of French. I disliked him immediately for the swagger with which he pushed the other Africans around and for his cowering and whining whenever he talked to us. Several park guards, whose duty it was to patrol the boundaries against poachers and wood cutters, appeared and built a shelter uphill from us.

In the ensuing days we attempted to study the gorillas in this typical bamboo habitat. Alone or together we searched the forests for chewed food remnants and for nests. We found some eighty-five nests of which about one third were on the ground, the rest in the tops of the bamboo some ten feet above the forest floor. We heard gorillas hooting and beating their chests in the distance, and we tracked them up and down the slopes without even glimpsing them in the maze of stems. We were always wet; nothing dried in the damp air. The trees dripped constantly, and every touch of a bamboo stem brought a shower of freezing drops. Our campsite was a quagmire. Yet, somehow, the moments of greatest misery ultimately accord the most pleasant memories. I remember one evening as Doc and I jogged along the winding paths made by elephants and squirmed through thickets of vines, completely soaked, trying to reach camp before darkness. Nearly exhausted, we reached our shelter, and Kay was there to hand us hot cocoa. We drank it gratefully and warmed ourselves by the fire until our pants steamed and the chill left our bodies. There were a few sunny mornings too, when the dew glistened on the leaves and the yellow-green bamboo shone with a soft velvety hue that made me want to reach out and stroke the hills. Shafts of sun slanted between the stems, by their very brightness emphasizing the subdued tone beneath the canopy. I liked to stand quietly and listen to the silence. Few birds call bamboo their home, and only occasional gusts of wind rustle the leaves overhead. Sometimes we climbed to the flat summit of the mountain and gazed into the rift valley, always shimmering in the sunshine even when we were standing in the rain.

Doc left us after a week to look for gorillas in the mountains to the south. The sky remained leaden, and Kay spent most of her time in our dreary camp. She prepared the meals and took photographs; she also had the thankless task of drying the blotters of the plant press one at a time over the open fire. Since I preferred to search for gorillas alone without having to argue with my guides about where to go and when to return home, the Africans in our camp had little to do. But Christoph was an excellent storyteller who acted out his rousing plots, even providing the sound effects. The others listened raptly to his stories, punctuating them with laughter or groans. The art of listening to a story has been almost lost in our society, except by children, but the natives who cannot read still have it. At other times they sang an endless song to which stanzas were added to fit the situation. Kieko wrote down several verses for Kay in his upcountry Swahili, which is much simpler and less grammatical than that spoken farther east:

Karisimbi Mikeno na Shabinyo ni mulima baridi mvua
Nyiragongo ni mulima wa moto

Walisema kiabirimu kuna ngagi tunachoka
Kuta futa ile ngagi kumbe iko bongo jawa garde

Heri mama heri kwetu heri kwetu
Wavi jana wa Rutshuru.

Karisimbi, Mikeno and Sabinio are mountains wet and cold,
Nyiragongo is a mountain of fire.
They say that one can see gorillas on Mt. Tshiaberimu,
We are tired of searching for those gorillas because they exist only
    in the imagination of the guards.
Peace to our mother, peace to our house, happiness to our house,
    those who were born in Rutshuru.

It has often been said that mountain gorillas live primarily in bamboo and prefer that vegetation to all others. In fact, bamboo covers only a small portion of the mountain gorilla's range. We found that in bamboo the main food of the apes consists of the tender white shoots by which this grass usually reproduces. Although we collected nineteen food plants on and around Mt. Tshiaberimu, bamboo shoots furnished the bulk of the gorilla's diet in the season we were there. Even the dung of the animals became soft, for shoots lack the fibrous material to bind it together. After checking various bamboo forests at all times of the year, it became apparent that the appearance of shoots is seasonal, with an abundance to be found at the height of the rainy seasons and often none at all during the driest months. When shoots are absent, gorillas must seek other forage, which may be scarce. All in all, the rather barren bamboo forests are not highly favored by gorillas.

I had not yet visited the northern part of the Mt. Tshiaberimu massif, and to do so I would have to spend a night on the trail. On May 25, I hiked northward through the fields and villages in order to save time before entering the forest on my journey back to camp. In a country where officials rarely leave their offices and still more infrequently step off the roads, I, as a lone white man with a pack on my back, presented an enigma to the villagers. They crowded into the doorways of their huts and watched me silently, for they were distrustful of all strangers. Children followed me giggling and laughing, daring each other to go closer. When I suddenly stopped and looked back at them, they screeched and fled, their slender brown legs whirling like windmills. Soon this became a game between us, and the ever-increasing horde of children ran at my heels over the hills until I disappeared into the bamboo.

Not long after entering the forest, I heard the *pok-pok-pok* of a silverbacked male beating his chest. I traced the route of the animals by the holes they dug in the earth to reach the base of the small bamboo shoots and by the broken young stems from which the ends had been chewed. The animals traveled rapidly, and I gave up trying to follow them. The sun was shining for once, and in a little grassy clearing on a ridge I reclined against the trunk of a tree and ate my lunch of crackers, cheese, chocolate, and dried fruit. Farther on, in another clearing, I came upon a snare set for the blue monkey (*Cercopithecus mitus stuhlmanni*), a beautiful creature with black feet and tail and a grey-black body ticked with white. An artificial bridge of bamboo had been constructed across a clearing, and a monkey tended to travel across it to avoid descending to the ground. But a bent pole with a snare was placed in such a way that the monkey had to enter the trap, and when it did, inadvertently pressing down a wooden stick, the snare was released with a jerk, strangling the animal.

After discovering this sign of poaching, I examined the trail ahead more carefully. A few days before a guard and I were hiking through the bamboo south of the park boundary and came to a slim liana stretched across the trail about four feet above ground. I was in the lead, and I stopped to look around, trying to determine the reason for it. Then I glanced up. Poised twenty feet above me in a tree was a six-foot log, weighing at least one hundred pounds, with a sharp iron spear in the end pointing wickedly at my head. It was an elephant trap. As the elephant wanders down the trail at night, it breaks the liana with its body, and the spear plummets into the back of the animal with the full force of the log behind it.

I was following the crest of a ridge along one of the many old elephant trails that crisscrossed the bamboo. Soon the tracks became fresh. The toenails were still clearly defined, and swarms of tiny black flies hovered about the heaps of dung. I pushed my fingers into some dung. It was still warm. Clouds drifted in, and grey fog crept from stem to stem, reducing my visibility to about fifty feet. I continued on silently and carefully, straining my senses, trying to hear the swish of a branch, the rumble of a stomach, trying to see the bulky grey forms of the elephants in this shadowless, dusky world, trying to smell their musky odor. But the only sound was the pounding of my heart. I was afraid of stumbling upon the herd, for it would be dangerous to have to elude them in this fog. Finally I talked to them in a normal voice: "Elephants, hello. Please get off the trail. This ridge leads toward my camp, and I don't want to leave it. I am only a human being, a weakling without weapons. I can do you no harm. Please leave the trail and let me pass." And just ahead, without a sound, the elephants left the trail and angled into the valley.

A drizzling rain soaked me. The ridge on which I was traveling divided into several lesser ones, and I crossed a valley or two, heading, I thought, in the direction of camp. As I slogged along, looking for a stand of trees under which I could build a fire and spend the night, I noted an oddly twisted bamboo stem that seemed familiar. I dropped to my knees and examined the ground and found my own footprints. From then on I trudged along compass in hand. Suddenly my feet stepped into air and I threw myself to one side, landing at the edge of a pitfall concealed in the trail. The pit was V-shaped and lined with bamboo poles so that any small animal, like a red forest duiker, is securely wedged in by its fall.

As I traced the edge of a marsh that nestled between rounded hills, I stumbled with great luck upon a small lean-to beneath some trees. The crude shelter, undoubtedly built by poachers, consisted of a sloping roof of bamboo stems covered with a layer of sedge. I collected some firewood and after much blowing the sodden branches blazed, creating a cheerful circle of light in this gloomy world. I set a can of beans into the flames and buried a potato in the coals. I hung my pants, jacket, and socks on sticks by the fire and crawled into my sleeping bag. Then, while my dinner was cooking, I whittled a bamboo spoon with which to eat. My wet clothes steamed, and the smoke hovered under the roof of the lean-to before pouring out and covering the marsh with a downy quilt. Some frogs croaked forlornly, and once a twig snapped. All around stood the stems of bamboo, their heads in the clouds. I ate my beans. The crisp skin of the potato opened with a steaming puff. After eating, I banked up the fire and lay listening on my bed of grass. I could imagine no more perfect evening than that.

We returned to Masereka on May 30 to meet Doc, not at all unhappy to leave our camp on the slopes of Mt. Tshiaberimu. Doc had had a highly successful week of mapping the distribution of the gorillas in this region. Today the gorilla survives only in a few isolated pockets, completely surrounded by cultivation. In 1928, when Mr. and Mrs. Hurlbert of the Kitsombiro Mission walked across these mountains, the forest was extensive and the people scarce. Now the reverse is true, and the future of the gorilla tenuous.

Doc also found that the gorillas occasionally leave the depth of the forest to raid maize and pea fields, a habit which does not endear them to the natives even though the amount of damage done is, on the whole, negligible. The gorillas must have acquired this habit in recent years, for maize was not grown in Uganda before the arrival of the Arabs in 1844 and probably not around Mt. Tshiaberimu much before 1920. Adult gorillas hesitate to try new foods, and I wondered how they had acquired the habit of eating maize and peas. Several gorillas at a research station in the Congo had been trapped as adults in the lowland rain forest where bamboo does not grow. When they were presented with the tender shoots of bamboo they refused to try them. Baumgartel tried to bait wild gorillas by putting out bananas, sugar cane, and other tempting morsels, but the gorillas refused to eat these foods unknown to them. Yet a large infant gorilla captured by Baumgartel's guides immediately devoured bread, pineapple, carrots, and other foods. Apparently young gorillas are more adventuresome in their tastes than adults. Dr. Imanishi noted that among Japanese macaque monkeys the female learns to eat new food items, such as peanuts and candy, from her own infants. Perhaps gorillas adapt similarly.

The Mt. Tshiaberimu region was also our first visit to an area where the gorillas are hunted by the natives for food, in the spirit of communal sport. Although the mountain gorilla has been completely protected by international agreement since 1933, there is no way to enforce such a rule in the remote villages; and, at any rate, natives always claim that they had to kill in self-defense. In various villages, Doc obtained the skulls or other remains of fourteen gorillas, indication of sizable slaughter. Between 1950 and 1959 the Kitsombiro Mission hospital treated nine injuries caused by gorillas. Three of these were minor, but the others required hospitalization. During a round-up of a lone silverbacked male, three Africans were bitten. Just before our visit to the area, a lone blackbacked male had wandered away from the forest, across the fields, and into a wattle grove. There he was surrounded by natives with spears. Infuriated, the gorilla lunged after an African, grabbed him by the knee and ankle and bit the outside of his calf, stripping off a piece of muscle seven to eight inches long. A nurse at the Kitsombiro hospital told Doc that the natives are sometimes bitten during a hunt when attempting to rescue their dogs. In all these instances man was the aggressor, and the gorilla was defending itself. Under such conditions, with all retreats cut off, even the most docile of creatures will attack, giving the impression of being ferocious. It is as George Bernard Shaw commented: "When a man wants to murder a tiger he calls it sport; when the tiger wants to murder him he calls it ferocity."

CHAPTER 5

A Walk in the Forest

While Sumaili scratched his head with a wooden five-pronged comb and the others built a fire, I ambled away into a valley heavily overgrown with brush. Slowly I pressed against the pliable wall of vegetation, the interlacing of lianas and thorny branches, and made my way into the thicket. A forest pig had excavated some roots, leaving a small crater and loosely scattered earth. Farther on lay the blue and white-spotted feather of a guinea hen. And later I found several branches stripped of bark by the dexterous hands of a gorilla. Nearby a cicada droned deeply then rose to a raucous screech. My aimless meanderings took me to the banks of a small stream. A brown snake glided slowly across the shade-dappled creek and disappeared in the opposite bank. I moved on against the current, and when I ducked beneath a bough, my neck suddenly burned like fire. Violently I slapped at and crushed the stinging ants that had dropped from the foliage onto my skin.

We drove southward on June 6 along the blue-green waters of Lake Kivu and over mist-shrouded mountain passes toward the town of Bukavu. Having finished our surveys of gorillas in the Virunga Volcanoes, the Impenetrable Forest, and Mt. Tshiaberimu, we now intended to concentrate our search in the vast uplands bordering Lake Tanganyika and in the rain forests of the Congo basin west of the rift escarpment. Bukavu is a beautiful town built on the promontories and coves of Lake

Kivu. Crimson-blossomed *Spathodea* trees shade the broad avenues and the shops carry the latest imports from Brussels, Paris, and Copenhagen. Some 5,000 Europeans had settled in the provincial capital of the Kivu, together with about 27,000 Africans, who lived mostly in housing developments and shanty towns in the surrounding hills.

About twenty-five miles north of Bukavu, on the slopes of the escarpment, lies the I.R.S.A.C. station (L'institut pour la Recherche Scientifique en Afrique Centrale), at which research in anthropology, protozoölogy, virology, botany, and other fields is carried out. The station helped us locate two Africans who were familiar with the gorilla country to the south, and we hired them for the journey. Although the cost of lodging at I.R.S.A.C. compared favorably with the most exclusive hotel in Bukavu, we decided that the station would be a fine place for Kay to stay while we spent two weeks surveying the forests.

We drove south, down the valley of the Ruzizi River. The fields of cotton had gone to seed, and the women and children were picking the white balls from the plants. The hills of the escarpment over which we later passed were steep and barren, thinly covered with grass and fields of talus. The valleys harbored a few villages and missions. At night we stayed at one of the many rest houses, or *gîtes*, which lie scattered over the country. The *gîtes* are simple houses, sparsely furnished with a bedstead, a table, and some chairs; they are designed to lodge government officials for a night.

The following day we were told at the Lemera Mission that high above us in the remaining patches of forest on the summit of the escarpment gorillas were believed to live. We found nests at one place, and later, with the son of one of the missionaries, I attempted to locate more spoor farther to the north. We passed on foot through several villages surrounded by banana groves. Here the dried manioc tubers were pounded into flour, the goats milked, and the chickens and children were running. Ahead of us we heard the incessant throbbing *bum-bum-bum* of a drum. We looked into the low-ceilinged room of a hut where bare-chested men squatted on the floor and clapped their hands, their white teeth shining in the dim light and their bodies glistening with sweat. In one corner a man was beating in frenzied tattoo the end of an iron barrel. A woman dressed in a loincloth gyrated on the earthen floor, arms swaying above her head, faster and faster, a wild ebony creature caught in the love of rhythm.

On June 15, we reached the town of Uvira at the northern end of Lake Tanganyika. In the midday heat the town was dead. Africans dressed in white gowns sat cross-legged on the verandas of the few stores. The road was deep in dust, and the feathery palm trees provided little shade. Uvira seemed to arrest time, and I did not find it difficult to imagine the explorer Speke walking down this street in 1858, the year that he and Burton discovered Lake Tanganyika. And I could see the gentle Livingstone and the cocky Stanley sitting beneath the mango trees that grew at the edge of the town after their famous meeting in Ujiji in 1871. In those days Uvira was a slave-collecting center. The Arabs who came in search of slaves in 1850 were the first outsiders to penetrate into Central Africa, and they led the way for such men as Speke and Stanley. Without the help of the slavers, many explorers could not have advanced very far. The Arctic explorer Stefansson somewhat cynically observed that a country is generally said to be discovered "when for the first time a white man — preferably an Englishman — sets foot upon it."

South of Uvira, the road was hemmed in on one side by the slopes of the escarpment and on the other by the blue waters of Lake Tanganyika. Large dugout canoes lay in quiet coves and masses of sardine-like *daaga* fish were spread to dry on the white sand beaches. There are nearly two hundred and fifty species of fish in Lake Tanganyika, many of them endemic. In contrast, Lake Kivu, which is geologically young and is guarded by turbulent rivers, has only about fifteen species. We had to drive carefully, for its was toward evening and the roads were crowded with Africans tipsy from the effects of *pombe* (banana beer). A hundred years ago Burton noted that *pombe* drinking started at dawn and continued throughout the day. In this respect at least times have not changed. There are a number of ways of making

Two lions at a buffalo kill in Queen Elizabeth National Park; hippopotami lie partially submerged in the water.

Dr. Emlen inquires about gorillas in a village in the lowland rain forest region.

Bamboo has many uses. The hut is built entirely of bamboo, and the African is weaving a basket of pliable young bamboo stems.

A female gorilla of group VII *with an infant about ten months old on her back.*

A village of mud huts surrounded by stands of bananas, some young secondary forest, and in the distance the mature lowland rain forest in the Congo basin.

pombe. One way is to bury the bananas in a pit covered with leaves and earth. After three days the earth is removed and after three more days the bananas are taken out and placed in a wooden trough. A man tramples the fruit to extract the juice. Some sorghum flour is added, and the mixture is then allowed to ferment for a day or two. The result is a pale whitish fluid, sour to the smell and the taste.

Near the shore of the lake we passed a cluster of small barren rocks named the New York Herald Islands. This was Henry Stanley's curious tribute to the newspaper that sponsored his African visit. Soon the rift mountains fell away from the lake, and we drove on through dry, rolling savannah.

On the map the town of Fizi, the Swahili word for hyena, looks like a thriving community. In reality there is, we soon found out, only a hospital, a guest house, and the office of the local "administrateur," the Belgian equivalent of the British district commissioner. To use a Congo map is always a sporting proposition. Villages may have moved or vanished, and roads, marked with a vigorous red line, may be impassable or abandoned, or yet to be built. We asked to be directed to a garage, for our car needed simple servicing, and were told that there was no garage in Fizi territory. The nearest one was in Usumbura, one hundred miles away.

To the west of Fizi, where the rift mountains drop sharply away to the savannah below, is the southernmost extension of the mountain gorilla range. One narrow mining road leads northward through the rift mountains, and this we took. We now entered a vast region of grasslands and mountain rain forest, and on the highest ridges were stands of bamboo. There were few villages and no towns; the only industries were mining and missions. The region was so immense that we could do no more than sample it. We resorted to second-hand information about gorillas from miners, government officials, and especially from the native residents. The road provided the principal means of foot travel for the Africans, and we encountered single individuals and small parties every few miles. Some were so shy that they dove into the bushes as our car approached, but most were willing to respond to our questions: "Are there gorillas on this mountain? Are there any in this direction? Can you show us any nests?"

Whenever possible, we checked their answers by searching for nests and other gorilla spoor, and our informants always proved to be correct. In this area the gorillas apparently cross short stretches of grassland from one forest island to another.

The rolling grasslands cease north of the Elila River, and the topography becomes rugged. Rain forest, similar to that found in the Impenetrable Forest, covers the terrain, except where the steepest slopes are broken by rock bluffs. These are the Itombwe mountains. Rudolf Grauer was the first European to penetrate this area in search of gorillas. In 1908 he spent three months here, collecting twelve gorillas for the Vienna Museum. The Wabembe in this region believe that the gorilla, the *kinguti*, is not an ape but a man who long ago retreated into the forest to avoid work. We saw no gorillas during our brief hikes into the forest, but we noted or were told of twenty-five kinds of plants which the apes ate, and we spotted their nests. But of the gorillas away from the road we learned nothing specific. One administrator pointed to a bare space on the map where the tentative courses of rivers were indicated by dashes. "It is — what do you call it? — no-man's-land. Nobody has been there; nobody lives there except perhaps a few Batwa."

We also stopped at the mining camps to ask for information. These mines have a tenuous foothold in the jungle. For five, perhaps ten years, the placers of gold and tin pay off and then are abandoned. If the roads which the mining companies have built are not taken over by the government they too revert to forest. It is a lonely life for Europeans in this region. Only the lure of high pay induces them to leave the crowded streets of Brussels and Antwerp. They are outsiders, and the everpresent forest is to them a closed world which they do not try to understand or approach. The vigorous life in the wilderness around them troubles their hearts, for their nerves are not attuned to the presence of things strange and unusual, to the threat of being alone with one's thoughts, or lacking the safety and organization of civilized surroundings.

We returned to the I.R.S.A.C. station on June 21, and five days later left for our last extensive gorilla survey in the Congo basin.

Between Bukavu and I.R.S.A.C. a highway winds up the rift escarpment. After dipping down into the Congo basin it continues on to Stanleyville on the banks of the Congo River some four hundred miles to the northwest. This road is but a narrow slash of yellow in the vast expanse of equatorial rain forest that stretches without interruption over a thousand miles to the west coast of Africa.

The Congo basin lay below us, first steeply undulating, then gently rolling, and finally flat. The cloudlike billows of foliage extended to the horizon, an unbroken quiltwork of varying shades of green. Once the basin had been an inland sea. Lying at an altitude of from one thousand to fifteen hundred feet, it is surrounded by mountains: the rift escarpment in the east, the Katanga highlands in the south, the Cristal Mountains in the west, and a ridge of land to the north, separating it from the Chad basin.

The road off the escarpment is steep and extremely narrow. Traffic is permitted in one direction three days a week, then three days in the other direction; on Sunday two-way traffic is allowed. This makes the road highly dangerous, for the Belgians tend to ignore the restrictions or do not keep track of the days of the week. As we descended, the crisp mountain air gave way to the hot, humid weather of the lowlands. Although the temperatures rarely climbed over 85°F, the high humidity made the air oppressive and enervating. The road evened out as soon as the mountains were behind us and followed the valleys and contours of the hills. My first impression of the lowland rain forest was one of flamboyance, a vegetation that had flung all restraint to the winds. An inextricable mass of vines, herbs, shrubs, and saplings crowded the sides of the road, and a little farther back stood the dark wall of the forest trees. Every few miles we passed a village of rectangular mud huts surrounded by groves of bananas, all but lifeless in the heat of the day. A few men lay in the shade, and chickens scratched in a desultory manner in the hardpacked earth. At the sound of the car only children raced to the edge of the road to see who was passing.

This was the Maniema Forest, the land of cannibals, the dark heart of Africa. Even the major explorers had shunned these endless jungles, and no road penetrated the area until World War I. The main slave trading route passed between Ujiji, Kabambara, Kasongo, and Nyangwe, far to the south. Nyangwe was established by the Arabs in about 1860 on the banks of the Lualaba River, as the upper Congo River is called. Livingstone visited Nyangwe in 1871, and in 1876 Stanley descended the Congo River to the sea. The intrepid Cameron, the second European to traverse the African continent, crossed the Ulindi River, a tributary of the Lualaba, in 1874. Emin Pasha, the German who very nearly became a sovereign over the northern Congo and southern Sudan, was assassinated by Arabs to the west of the town of Lubutu in 1892. Not until 1894, when Count von Götzen descended the Lowa River, another tributary of the Lualaba, did an explorer actually penetrate the heart of the Maniema Forest.

The Warega, the Wanianga, and other tribes in the area are Bantu-speaking agriculturalists. They were cannibals, an activity which the Belgians tried to stamp out. The last known outbreak occurred in 1958. The men of one tribe captured men of another tribe and sold them — as meat on the hoof, so to speak — to a third. The way of life of the tribes is largely determined by their method of agriculture, which is of the slash-and-burn type. With ax and fire they fell the forest trees, which then decay on the ground. After the smaller trees and shrubs have been burned, banana shoots are pushed into the ashes and shallow soil, and manioc may be planted beneath the bananas. Sometimes mountain rice, a grain which can be grown like maize on hillsides, is planted first and other crops planted after the harvest. But the soil which can support some of the most luxuriant plant life on earth is soon exhausted by cultivation. The humus decomposes rapidly in the direct rays of the equatorial sun, and the mineral content is leached out by the heavy rains. After three or four years the field is exhausted and must lie fallow for at least twelve years before a new crop can be grown. More forest is cleared and more bananas are planted, and villages are moved to be nearer the fields — the endless, ageless cycle of shifting cultivation.

In the abandoned fields the new forest rises phoenix-like from the heart of the ashes. At first grasses and vines cover the soil, but after a year or two shrubs and saplings have pierced through the grass to win the open sky. After the plot has been abandoned for five or ten years, the vegetation consists of an impenetrable tangle above which towers the *Musanga* tree, its cluster of shiny leaves radiating outward from the petiole like an umbrella. This tree has a phenomenal growth, reaching a height of forty feet in five years, only to die of old age at about twenty years. As the new or secondary forest grows and as the canopy becomes more and more continuous, less and less sunlight reaches the ground. The plants in the jungle can be successful either by struggling upward into the sunlit canopy or by adapting their needs to life in the gloom of the forest floor. Many of the colonizing plants cannot adapt; as the forest grows older, the undergrowth disappears almost entirely. After about eighty years, tall forest again covers the ground where once villages and fields stood. The constant cutting, planting, and abandoning have created a landscape consisting of ephemeral fields surrounded by a patchwork of forest in various stages of regeneration, and always on the horizon, the overpowering presence of the rain forest.

The villages are but islands, mere footholds, in the jungle. The forest world beyond the fields is to the people a strange and fearful place which they do not often enter. Their life centers on their fields where they wage an endless battle against the ever-encroaching jungle. As if to eliminate the forest from consciousness, all trees are felled near the villages, depriving them of shade. Although the natives have lived in the forest for perhaps two thousand years, they have remained outsiders. At the end of the Pleistocene, Negroes were probably living in the grasslands and savannahs of West Africa, and only pygmies roamed the forests. Sometime before 3000 B.C. agriculture arrived from Egypt, and

Negroes domesticated millet and sorghum, plants that do not thrive in rain forest. Later they also cultivated yams, which, together with iron that was introduced just before the Christian era, enabled them to invade the forest. Bananas were not brought in much before 1600, and manioc not until around 1750.

After several hours, we reached the mining camp of Kabunga, one of many mines in the area run by the C.N.Ki. (Comité National du Kivu) which in 1928 received exclusive mineral rights to three-quarters of a million acres of forest. Not far from Kabunga, in a forest clearing, was the home of Charles Cordier and his wife. We had been told that Cordier, an animal trapper of Swiss extraction, knew more about the gorillas in this region than anyone else, and we had come to ask him for information.

Charles is a tall, robust man of over sixty with grey hair and a youthful face. He is a man of violent loves and violent hates, and his genial disposition is masked by his gruff voice, yet I have rarely met a man whom I have admired more. There are many kinds of collectors who supply the zoos with animals. Most make quick journeys to some far-off land, catch and buy anything that moves, and return with a grabbag of miscellaneous creatures about which they write a book. Charles, on the other hand, concentrates his energy on only the rarest and most elusive animals, those that have never been exhibited in zoos before, or those that are exceedingly difficult to trap. Years may be spent in pursuit of a single species.

He has a great liking for and understanding of the animals he handles, traits often sadly lacking in collectors. In West Africa, for example, an unscrupulous American dealer shot down whole gorilla groups in order to obtain the infants. Most of these infants died from diseases — or sheer loneliness — before reaching zoos. Here is an extract from a report by a collector for a medical institute:

On the other side of a small clearing a female was playing with a small baby. Everything seemed perfect for a good shot. The gorilla wasn't fifty yards away and was unconscious of our presence. But I couldn't help thinking of the other gorillas that we could hear, and couldn't see. . . . But there was nothing for it, I had to shoot. I took plenty of time and when I stopped shaking, I made a clean hit through the skull, killing her instantly. [The male rushed up.] I fired and hit him in the shoulder. He staggered for a second, but kept going. I fired again, and again he staggered. . . .

The infant, so mercilessly obtained, died a few days later because no provision had been made for its care. For each of the eighty-five gorillas alive in the United States today, at least five others died while being captured or before they reached a zoo, a sad commentary not only on many collectors, but also on zoos, which for the most part care little how their animals are obtained.

Charles Cordier, on the other hand, developed a technique of surrounding whole gorilla groups with nets. One hundred reluctant natives are required to hold up the nets into which the apes charge, and then are captured. This is an expensive and time-consuming method, but it is humane. Unwanted animals are released. A mountain gorilla will cost you $5,000, which at first seems a high price. But when all the effort, time, and expense of capturing and transporting the animal to a zoo are added, it is reasonable enough.

The life of Charles Cordier has been a hard one, for he is one of those spirits who will endure little compromise. Yet he has never lost his curiosity and enjoyment of the world around him. In 1959 and 1960, the *Kakundakari* was the focus of his enthusiasm. The Kakundakari is to the Congo basin what the Abominable Snowman is to the Himalayas. According to the natives the male is five and a half feet tall, the female four feet. Their bodies are covered with hair, and they walk upright. At night they sleep on beds of leaves in caves, and during the day they forage for crabs, snails, and birds. Charles claims that he saw a footprint of these manlike creatures, and one was supposedly killed in 1957 at a mining camp. Another stumbled into one of Charles's bird snares, fell on its face, turned over, sat up, took the noose off its feet, and walked away before the African nearby could do anything. Is the Kakundakari a gorilla, an ape-man, or a myth? Can large, unknown creatures still exist undiscovered in Africa?

As recently as 1901 Sir Frederick Jackson wrote the following letter to a collector:

SIR:

"I am directed by H.M. Special Commissioner to call your very special attention to the fact that an extraordinary animal inhabits the forests in the Mboga subdistrict of Toro, and to urge you to do your utmost to procure perfect specimens of the same.

The animal in question appears to be a connecting link between the antelopes and giraffes. It is known to the natives by the name "Okapi" . . .

The mountain gorilla emerged from the forests of Africa in 1902, the giant forest hog in 1903, the Congo peacock in 1936. Thousands of square miles of the Congo basin remain uninhabited by man and are unexplored. At present I find no reason to deny the existence of the Kakundakari, and I hope that some day Charles will find the creature.

Charles showed us his animal collection, which was housed in a long low shed. There was a pair of Hartlaub's ducks, dusky brown with a splash of light blue on the wings, shy birds of quiet forest pools; there were fruit pigeons, ibises, grey parrots; and there were Congo peacocks, which Charles was the first to capture. Congo peacocks are very difficult to find, for the male calls only at night, a loud *ko-ko-wa*, and the female answers *hi-ho, hi-ho*. Small flocks wander down the forest paths in single file with the male in front, and ten males for every female are thus caught in the snares. There were also giant pangolins, their body covered with overlapping scales, giving them the appearance of fish lost on land. Somewhat removed from the house, at the edge of the forest, stood a high stockade which contained "Munidi," a magnificent silverbacked male gorilla, an adult when Charles trapped it.

Two infant gorillas shared the house with Charles and Emy. They were demanding charges, needing as much love and attention as human babies. Noël, so named because she was obtained on Christmas Day, slept in a crib in the Cordier bedroom. She was a shy creature, less than a year old, who held Emy tightly in the presence of strangers. On the screened front porch lived Mugisi, a rambunctious two-year-old. Denis, an African boy, was hired to entertain this youngster throughout the day, and he obviously took pleasure in his task. When he went home for the night, the watchman had to baby-sit with the ape until it climbed into its nest box, dragging along its favorite cloth. Mugisi was a tease and a show-off, a happy and uninhibited youngster, who, after a period of initial restraint, became good friends with Kay and showed his liking by trying to bowl her over by charging into her legs from behind.

Charles pointed out to us on a map where he knew we would find gorillas or where he had heard rumors of their presence. Soon Doc and I realized that we would have to work independently in order to cover some fifteen thousand square miles of forest in a reasonably short period of time. Doc would inquire about the presence or absence of gorillas along the roads, at missions, and at mining camps as we had done in the Itombwe Mountains. I had become intrigued by an uninhabited stretch of forest that lay to the south of us. There were no villages, no roads, and I knew of only one European, the geologist André Meyer, who had penetrated this area in recent years. If I hiked some sixty-air-miles south from Utu, a mining camp just to the west of the Cordier home, I would reach another highway where Doc could pick me up. The actual distance on the ground would be close to eighty miles, and we figured that I should be able to do this in four days if I traveled light. Charles kindly lent me Sumaili for the trip. Sumaili's muscles bulged beneath his tight black shirt and his incisor teeth were filed to sharp points. He disdained *pombe*, a fact so unusual that several Africans made a point of mentioning this to me. We tried to find another African to accompany us, but this was a difficult task, for no one seemed eager to make the trip, let alone with a European. At length we located two men in one village who agreed to come with us for 300 francs each, and we arranged to meet on the morning of June 29.

At dawn the four of us stood on the road at the edge of the forest and arranged our loads, dividing the weight equally among us. Wisps of fog crept over the ground, and far away a rooster heralded the day. I had brought a minimum amount of equipment — a jungle hammock, a sleeping bag, a camera, binoculars, a small pot, one spoon, a tooth brush, and food enough for six days. Bakire, a wiry fellow with a ready grin and a heavy growth of whiskers, wrapped his load of rice, bananas, and a change of clothes into a blanket. He cut a sapling with his machete, tore off a long strip of bark with his teeth, and tied his blanket with it, leaving a loop to serve as tumpline. Mtwari carried a small suitcase which contained his shoes, clothes, and other odds and ends; Sumaili had his gear and some of mine bundled into a cloth.

In single file we stepped into the dark, soft atmosphere of the rain forest. The black trunks of the giant trees were spectral in the morning mist. Our voices sounded like noisy interruptions in the immense quiet, and soon we spoke no more. Undisturbed rain forest is kind to intruders, for the vegetation is relatively scant near ground level. We wandered freely between the trees on a carpet of sodden leaves. Lowland rain forest closely resembles that in the mountains, except that here the trees are larger and the hanging shapes of lianas are more common. The first rays of the morning sun suddenly suffused the forest with a soft rust-red glow. As we trudged along in the strangely silent jungle there was only the drip, drip of the dew and occasionally the hesitant call of a bird. Once a small band of red colobus monkeys took flight high above the canopy, but we saw little more than the swaying of branches and the shadowy forms of the long-tailed creatures.

We were heading straight toward the south, sometimes following what seemed to be a foot path, at other times across ridges and through valleys without the sign of a trail. We crossed and recrossed nameless streams and, when the brush along the banks was dense, we waded down the middle. Our tracks mingled with those of buffalo and, rarely, elephant. We came across the hoof prints of a bongo, a rare and beautiful forest antelope with vertical white stripes on its light brown coat. But the game in the forest is elusive, and we did not see any of these large animals.

The slight morning chill gave way to the midday heat. The air was still and waterlogged. The Africans wore only shorts, and their dark bodies glistened.

We stopped briefly while the Africans ate their mid-morning meal of boiled rice. They dipped their fingers into the communal pot, made a pasty ball, and popped it into the mouth. We never quite managed to have meals at the same time, for they ate again in mid-afternoon and in the evening. I adapted myself to their times except for breakfast, which I continued to make at dawn. Silently, and to our surprise, three men and a woman suddenly appeared ahead of us, bowed under the weight of baskets filled with rice. I found out that we were on a vague trail that connected the southern and northern villages, an ancient trading route used long before the coming of the first roads. In confirmation of this we occasionally came on crude lean-tos, constructed of branches and covered with paper-like leaves of the herb *Megaphynium*. These large ovate leaves are wonderfully versatile. With a simple twist they can be shaped into a cone — a handy disposable cup for quenching the thirst at the stream's edge. They can also serve as plates for rice and meat, as wrapping paper for small objects, and as roof shingles.

We decided to camp by mid-afternoon, for I wished to putter at leisure around the forest, and the Africans did not want to travel for more than six hours a day. While Sumaili scratched his head with a wooden five-pronged comb and the others built a fire, I ambled away into a valley heavily overgrown with brush. Slowly I pressed against the pliable wall of vegetation, the interlacing of lianas and thorny branches, and made my way into the thicket. A forest pig had excavated some roots, leaving a small crater and loosely scattered earth. Farther on lay a blue and white-spotted feather of a guinea hen. And later I found several branches stripped of bark by the dexterous hands of a gorilla. Nearby a cicada droned deeply then rose to a raucous screech. My aimless meanderings took me to the banks of a small stream. Low branches stretched peacefully toward the water in sweeping curves. Slowly I lowered myself into the stream, and the coolness of the water felt good against my weary legs. I drank deeply, secure in the knowledge that no village had contaminated the water. There I stood still, wholly relaxed but receptive to any small adventure that might come my way. A brown snake glided slowly across the shade-dappled creek and disappeared in the opposite bank. What kind was it, I wondered. I moved on against the current, and as I ducked beneath a bough, my neck suddenly burned like fire. Violently I slapped at and crushed the stinging ants that had dropped from the foliage onto my skin. On a damp ridge I caught a pinkish-white land crab. It had a small lid on its abdomen, and when I lifted it, I was delighted to find about a dozen miniature crabs with red legs, perfectly formed, crawling around in the chamber. Carefully I placed the crustacean on the ground and watched it scurry under a log.

I returned to camp, stretched my jungle hammock between two trees, and then joined the others by the fire. I pushed my pot of water into the hot coals and added some rice and a touch of salt; I opened a can of beef stew and heated that too. When the rice was cooked, I poured some of the meat over the rice and gave the rest to the Africans. We squatted in the glow of the fire and ate as the sudden tropical darkness descended and the world grew smaller until it consisted only of the circle of light cast by the flames of our fire. Our talk was desultory, about our present circumstances; the affairs outside the forest had lessened in our minds. Then we stared into the flames, eyes wide open, the forest quiet and motionless about us.

Mtwari had brought a lantern which he kept burning brightly all night. When I asked him for the reason, he replied: "Oh, Bwana, the forest at night is full of trouble, full of many dangers."

We rose to the dew-soaked forest of dawn. Shivering, I pulled on my wet and clammy clothes. And, as yesterday, we plodded along an indeterminable number of ridges and through broad valleys. At first the forest gave the impression of having been there since the beginning of time, a mysterious world in which the animals still ruled and man was a fleeting intruder. But a closer look revealed the destructive hand of man, even though his huts had long ago returned to the soil and the forest had reclaimed his fields. Many of the valleys were covered with shrubby forest, indicating that the tall trees had once been felled. Here and there were rows of palms that once had lined village streets. Many years ago this whole forest had been a patchwork of fields. What had happened to the people? The slave traders had decimated the villages, and so had such diseases as smallpox. When the Belgians attempted to establish their control over this area and tried to eliminate the constant tribal wars, they found themselves powerless in the immensity of the forest. By about 1920, the government had moved all Africans from the hinterlands to the roads where they could be handled more easily. Today, forty years later, the forest remains uninhabited, but tribal wars still flare up. A missionary told Doc that in 1944 the natives in this area clashed. Government soldiers killed six hundred of them before peace was restored.

I found surprisingly little sign of gorillas. Occasionally I came across the shredded herb *Aframomum* from which the apes had extracted the tender pith. In the soft sand along the banks of a stream we found the twelve-inch-long tracks of a lone male. We spotted nests too, but less frequently than in the other areas we had visited. Once we heard gorillas in the dense tangle of a valley. When I attempted to move closer, Bakire took my arm and shook his head, fear on his face. After a moment of hesitation I gave in, deciding that visibility was too limited to obtain much information.

That evening after we had camped, the Africans left to search for firewood. Soon they were back, dragging a dead forest pig, whose foot had been trapped by a natural snare of lianas. The small animal had died several days ago — its body was bloated, its skin greyish-green, and its odor absolutely vile. The voices of the Africans lifted in eager anticipation as they hacked at the carcass with their machetes. Most of the meat they impaled on sharpened sticks, which were set upright by the fire; other morsels were wrapped in leaves and laid near the hot coals. Mtwari hauled out the intestine hand over hand, cut it into sections, squeezed out the fecal matter, and stuffed the intestine into a pot of boiling water. The other entrails followed. Finally only the head remained. Sumaili smashed the skull and poured the brain into the stew, followed by the snout and beady eyes. They invited me to dinner, and laughed when I declined. I had tinned corned beef instead. The Africans gorged themselves far into the night and, half-asleep in my hammock, I heard them talking above the roar of a tropic rain that lasted for several hours.

The following morning, after the mists had cleared, the air was fresh, scoured clean by the rain. The streams had risen two feet, the homogeneous floods of amber water racing toward the Lualaba River. We came to a stream some eighty feet wide. Angry waters roiled by; sometimes a tree leapt from the depth and vanished again. A small log bridged the river. Nimbly the Africans scurried along the swaying bole, loads balanced easily on their heads. I humped along on my belly like a worm, grasping the tree securely with arms and legs.

"Tomorrow we will reach a village," Sumaili told me.

I remember little of that day or of the next one, except the heat and my weariness. The hills grew lower and the valleys broader, and suddenly in the evening of the fourth day, we stepped out of the forest and into a field. Soon we entered the most wretched village I had ever seen. There were six tattered huts, three on each side of a central courtyard. We entered one of these huts, empty except for a log which served as a bench, and a smoldering fire on the dirt floor. Everyone crowded in to watch me unpack. All eyes were riveted on me as I put some rice into my pot and asked for some water. The women giggled among themselves, and the children peered wide-eyed from behind their mothers. They were sad little creatures, naked, with bloated bellies, and eyes that were caked with a whitish mucus. One infant began to scream, and the others joined in. Soon only the men squatted around the fire, smoking coarse chunks of tobacco wrapped in bits of newsprint and talking in low tones. There were periods of utter silence when everyone stared at the floor, dribbling spit from their lips and grinding it slowly into the earth with their heels. A woman entered with a gourd of banana beer which made its rounds. Another brought a pot of boiled rice, and all dug in with their hands. In one corner of the hut huddled a husky fellow with the vacant stare and twitching movements often seen in the mentally deficient.

Hovering around the fire were four dogs, terrier-sized, yellow-black, with pointed ears and shifty eyes. They competed with several chickens for the grains of rice that fell to the floor. One pup, terribly emaciated, ventured too close to the fire and received a kick which sent it whimpering and snapping into a corner; another was swatted with a flaming brand. The dogs were continuously abused by the villagers. The whole grief of these animals was in their eyes, as they snuffed longingly around the huts, slaves of men, yet unwilling to find their way back to the wilderness and freedom of the nearby forest. The lack of compassion for animals in those African tribes which I came to know was very striking. They were also the only peoples I had ever met who never attempted to keep wild animals as pets.

No other trees grew near the huts, so I slung my hammock between two banana trees. Sumaili held a lantern high while I undressed, and the whole village crowded around to watch me retire for the night. The light of the moon played between the banana stems, and the faint tune of a *likembe* drifted from a hut as I fell asleep. A thunderclap woke me, and a violent gust of wind shook my hammock. After a brief pause, gray plunging rain, solid sheets of water, slanted down. I dozed intermittently and woke with a start when I found myself lying on the ground. The rain had loosened the earth around the bananas,which then slanted inward under my weight. I made my way to the nearest hut and bedded down in a corner. Several men woke up and the dogs growled, but soon all went back to sleep, the dogs huddled against the softness of my sleeping bag.

I rose with the light. Several chickens and I ventured together into the sodden world. Muscovy ducks splashed in the rivulets. Smoke rose lazily through the grass roofs. The lovely leaves of the bananas sparkled. In a doorway a woman stretched and yawned. Everything was very quiet, as though exhausted by tears. Mists obscured the forests beyond. We were a ghost island floating unchanged through time.

That afternoon we reached the mining camp of Niabembe, where at a rest house I began my wait for Doc. I was unshaven and dressed in filthy clothes, without having the means to improve my appearance. Sumaili, on the other hand, sported clean shirt and pants, and he had even carried polish and a shoe brush to make himself look respectable. All that afternoon the Africans filed past to stare at me.

When Doc and I compared notes after our survey, we found that we had extended considerably the known range of the mountain gorilla. Some three-quarters of all mountain gorillas inhabit not the mountains, as their common name implies, but the hot, humid Congo basin at altitudes of less than five thousand feet. A considerable number are found in one block of forest through the middle of which I had walked in the past few days. We noted that the apes were not spread evenly through the forest but that many were concentrated along roads and around villages. Gorillas favor those parts of the forest which provide abundant forage on or near the ground level. Under the canopy of the tall trees they find little to eat and consequently spend much of their time in the more open valleys and along rivers where, in the sunshine, the undergrowth is dense. Especially suitable for gorillas are the abandoned fields near the villages, for there the apes find their favorite foods in greatest abundance — the fern *Marattia*, the herbs *Aframomum* and *Palisota*, the fruit and leaves of the trees *Musanga*, *Myrianthus*, and *Ficus*. Shifting cultivation has been practiced in the lowlands for centuries, creating the kaleidoscopic pattern of forest regeneration so well liked by gorillas. Man has here a curious role in the ecology of the gorilla. He is both an enemy in that he hunts the animal for food, and he is also inadvertently the ape's benefactor by providing it with optimum conditions for living.

Sometimes the gorillas raid banana fields, eating not the fruit but the pith of the stem, ruining the trees of the Africans whose plantations they devastate. The natives band together, surround a group of animals and net, spear, and slash at anything that moves. At such times, Charles Cordier told me, the male after charging repeatedly, usually flees or is killed. The plight of the females and infants is described by the West African hunter Fred Merfield: "I have seen the native hunters, having dispatched the Old Man, surround the females and beat them over the head with sticks. They don't even try to get away, and it is most pitiful to see them putting their arms over their heads to ward off the blows, making no attempt at retaliation." Although attacks are usually made by males, females may also do so on occasion. Baumgartel related that a female placed her hands around the throat of his guide as if to choke him. Attacks usually consist of a lunge forward, a brief contact during which the gorilla bites, then retreat. There is apparently no prolonged mauling, and the wounds, although often serious, are usually not fatal. Many observers have noted that a gorilla often will not attack a person who stands his ground and faces the advancing animals. But as soon as the man turns and flees, the gorilla pursues on all fours and, like a dog, bites. Among the Mendjim Mey, a tribe in the Cameroun, it is a disgrace to be injured by a gorilla for everyone believes that the gorilla would not have attacked if the man had not fled.

Considering the constant hostility between man and gorilla in this region, it is not surprising that seemingly unprovoked attacks by the ape may occur. Between 1938 and 1940, in one district alone, fifteen Africans reported injuries caused by gorillas. But I suspect that most of these were obtained during hunts or when wood gatherers or field workers bumbled accidentally into a resting group of gorillas. Once, for example, I came upon two women grubbing for manioc tubers in a densely overgrown field. They were unaware that gorillas were foraging nearby. Even though gorillas are frequently pursued, they return again and again to the same general area. I have seen nests in trees within a hundred feet of inhabited huts. Tree nests were much more common in this area than in the other forests we studied, and I wondered if perhaps the frequent meetings with man may have been one cause of it.

All in all, and in spite of the mutual attacks, the primary responses between gorilla and man consist of respect and avoidance. The gorilla is by nature reserved and shy and, whenever it can possibly do so, it avoids contact with its human neighbors. Unless fully armed, banded together, and fortified with *pombe*, the Africans fear the strength of the gorilla. Once, in August, 1960, I followed a meek little man whom I had hired as a guide near Miya, a mining camp west of Utu. Suddenly branches crackled in the undergrowth ahead. My guide ran back and frantically motioned me to retreat. A female gorilla ascended a tree about a hundred feet to my right. Carefully I climbed into another tree and watched her. The group was well spread out, and I caught only occasional glimpses of an animal in the shrubbery. Another female climbed a *Myrianthus* tree and placidly plucked and bit into a ripe yellow fruit. Unfortunately, one female spotted me and with a high-pitched scream half-climbed, half-jumped into the undergrowth. The group congregated and watched me from the cover of the vegetation for a few minutes, then continued to feed. When I descended, my guide was nowhere in sight. On the way back to the car, nearly a mile from the gorillas, I found the guide clutching the upper branches of a tree, peering down at me with large eyes.

Yet in the end, according to native legend, the cunning of man prevails over the strongest brute. Once, long ago, people planted a fine grove of bananas. A large gorilla saw a woman harvest the fruit, rushed up, seized her and the bananas, and took them to his den. As he sat and ate, the woman screamed, and the gorilla squeezed the life out of her. On the following morning the gorilla was hungry again and returned to the plantation. He entered five huts and squeezed all the people to death. Only a boy and his mother survived, and they hid in the rafters of a hut. The gorilla brought several loads of bananas into this hut. "Ha," he said, "plenty of bananas and all my own." "Ha," echoed the boy in the rafters. The gorilla searched for the sound but could not find its source. Whenever he looked away, the boy stole the bananas and gave them to his mother. This went on for several days as the gorilla brought in more and more bananas, only to lose them. The gorilla grew very hungry, and his rages grew violent when he could not find the mysterious "Ha" that filled the room whenever he tried to eat. Finally his belly hung like an empty sack and he fell over and died from weakness. (H. Stanley, 1893, *My dark companions and their strange stories.*)

Although the Africans in the lowlands kill many gorillas yearly, their primitive weapons limit the extent of the slaughter, and the apes seem to be holding their own. I only hope that possible future improvements in agricultural methods will not disrupt the pattern of shifting cultivation, and that killing rates will not increase with the import of firearms. One reliable authority told me that in about 1948 officials organized the killing of some sixty mountain gorillas near Angumu to obtain eleven infants for zoos. One mining official near Utu bragged to me of having shot nine of the strictly protected animals for sport. These are isolated instances, but if they were multiplied by every meat-hungry African, the future of the gorilla would be most uncertain.

We also found that concentrations of gorillas may occur at such physical barriers to the forest-dwelling apes as extensive cultivation, grasslands, and, importantly, broad rivers. Gorillas apparently do not swim, and they even hesitate to enter shallow water. Charles Cordier told us of a group which refused to cross a stream less than two feet deep even when pursued by natives with spears and nets. Several times Africans showed me natural log bridges which gorillas regularly used to get from one bank to another. The only means by which gorillas can reach the other side of a large river is at the headwaters where they can cross on trees that have fallen to form a natural bridge. We found some confirmation of this at the Lugulu River. At one place this stream was two hundred and fifty feet wide, and it broadened still more before reaching the Ulindi River, a tributary to the Lualaba. No gorillas were known to occur on the southern bank of the river, but fifty miles upstream, where the river was narrow and strewn with boulders, gorillas had colonized the other bank.

Doc noted during his road survey near the town of Lubutu, in the northwestern corner of the gorilla's range, that isolated groups of apes existed in the vast expanse of the rain forest. Such pockets of animals, which we had already noted in the Itombwe Mountains, puzzled us, for no major rivers hemmed them in, the forest was continuous, and the altitude and topography were similar to other areas. Stragglers, lone gorilla males or small groups, are sometimes found twenty or more miles from the nearest large concentration of gorillas. Such erratically wandering animals are perhaps the principal means by which new territory is colonized. A small group may settle down far from a neighboring gorilla population and form the nucleus of a new one. It is also possible that once the gorillas were more widespread in these forests, and that the isolated populations are mere remnants.

On July 10 we returned to Rumangabo, our surveys essentially completed. After Doc left for home on July 16, Kay and I spent another month in making brief surveys of areas we had not covered. We had discovered that the mountain gorilla was more widespread and more abundant than previously supposed. Although its range spreads over some thirty-five thousand square miles of forest, the animals themselves are concentrated in about eight to nine thousand square miles of terrain. How many mountain gorillas are there in existence today? Previous guesses fell usually between one to two thousand animals. Our estimates are naturally tentative, for accurate counts of gorillas in a forest environment are difficult to make. After intensive surveys we found that in the Virunga Volcanoes there were roughly three animals per square mile, in the Impenetrable Forest less than two per square mile. Yet the relative frequency with which we encountered gorillas or their spoor in these two isolated and protected areas was greater than in the other forests we investigated. The population density for the mountain gorilla range as a whole probably does not exceed one animal per square mile, giving a total population figure of about eight to nine thousand animals. Until further work is done, I would estimate five to fifteen thousand gorillas — not enough to fill an average baseball stadium. History has shown that animals as rare as the mountain gorilla are highly vulnerable, and constant vigilance must be maintained to prevent the scales from tipping from security to extinction.

CHAPTER 6

Among Gorillas

On the last day of August, as N'sekanabo and I clambered up the boulder-strewn depths of Kanyamagufa Canyon, a tremendous roar filled the chasm and bounding from wall to wall descended the mountain like the rumbling of an avalanche. We started and ducked and then peered up at the silverbacked male who, surrounded by his group, stood motionless at the canyon rim looking down at us. Quietly and as unobtrusively as possible we retraced our steps under the watchful eye of the male, feeling chastised like children for having so crassly intruded into his domain.

The porters were waiting for us at Kibumba, at the mud-walled schoolhouse from which we always started our trek to Kabara. There were fifty-five porters ready to haul our supplies into the mountains, and ten others needed by Maurice Heine, a botanist from Rwindi, who accompanied us for the first part of the journey. Heine intended to spend three days climbing Mt. Karisimbi to check the rain gauge on the summit. We had brought enough food to last five months, a period during which we hoped to be independent of the outside world. We loaded up the porters with cases of tinned meats, fruits, and vegetables, with a sack of potatoes and bags of flour, with a basket full of live chickens and a bag of maize with which to feed these birds, with heavy clothing, bedding, and rain gear, with books and first-aid materials. The list of supplies seemed endless, but each item was needed, or at least might be needed. We had hired Andrea Batinihirwa, a twenty-three-year-old African, to do such chores as chopping firewood and washing clothes. A park guard, to be replaced about every three weeks, was to remain with us at Kabara to act as messenger in case of mishap, to accompany us into the forest when needed, and to serve as companion to Andrea.

The scraggly line of porters wound its way upward, tracing the edge of Kanyamagufa Canyon, up through the zone of bamboo and into the *Hagenia* woodland, over the same route which Doc and I had followed six months before. Recent rains had turned the trail into a morass, further churned by the hooves of buffalo and the feet of elephant. A downpour soaked the forest and wet our clothes, giving notice of the usual weather in these mountains. We were relieved to reach Kabara. While the porters huddled around an open fire in the shed trying to warm themselves before returning to Kibumba, Kay and I waded somewhat dejectedly among our sodden stacks of boxes, duffels, and other gear scattered in utter confusion around the hut. Our first task was to unpack, to turn this crude shack into a home.

We used one of the rooms for storage, stacking our food and spare equipment along the walls. A second room became our bedroom. Here we draped the walls with grass mats and gaily printed native cloth, not only to add a touch of color but to help keep out the wind that whistled through the cracks. A small table was covered with oilcloth and on that we placed a yellow water pitcher and a turquoise wash basin, further brightening the gloomy interior. The central room of the hut, and the only one with a door leading outside, became our main living quarters. In one corner, to the left of the tiny iron stove, Kay hung her pots and pans and stacked her dishes. It became her kitchen, where she prepared our meals, cooking either on the wood stove or on two small kerosene burners we carried with us. In the other corner was my office, as well as our living room, where we kept a typewriter, books, notes, and letters. As in the bedroom, we draped grass mats and cloth on the walls and tacked up maps of the Congo and of the world.

On the following day, while Kay continued to put our hut in order, I went into the forest to look for gorillas. It was the kind of morning that lifted my spirits. The air was crisp, and the sun sparkled on dew-soaked foliage. A white cloud covered Mt. Karisimbi. It poured down the slopes like a snowy avalanche only to be caught by the updrafts and danced skyward again, a tattered wall of ghostly flame. I headed northward over the rolling terrain toward Bishitsi. The forest was very quiet except for the twittering song of the southern double-collared sunbird. These warbler-sized birds, flitting about the foliage, gathering nectar from a lobelia flower, catching some insect on the wing, were shimmering, iridescent jewels. The upper parts of the male are a metallic blue-green, a broad band of red crosses his chest, and his sides are yellow. At intervals I found the pear-shaped nest of a sunbird hanging suspended from the tip of a swaying bough, a fragile structure of grass with only a small hole leading to the dark interior.

It was not easy to return to the wilderness after having lived in the villages and towns and traveled along man-made roads. I felt somewhat like a prisoner who, liberated after many years, does not know how to use his freedom and his faculties. In civilization one loses the aptitude for stillness, the habit of moving gently. It takes time to cease to be an outsider, an intruder, and be accepted once again by the creatures of the forest. The return to the wilderness is a gradual process, unconscious for the most part. Once the senses have been relieved of the incessant noise and other irrelevant stimuli that are part of our civilization, cleaned, so to speak, by the tranquillity of the mountains, the sights, sounds, and smells of the environment become meaningful again. Slowly the courage and confidence of man, previously nurtured by his belief in the safety of his civilized surroundings, slips away. Finally he stands there, a rather weak and humble creature who has not come to disturb and subdue but to nod to the forest in fellowship and to claim kinship to the gorilla and the sunbird. I recalled a Navaho chant:

> The mountain, I became part of it . . .
> The herbs, the fir tree,

I became part of it.
The morning mists,
The clouds, the gathering waters,
I became part of it.
The sun that sweeps across the earth,
I became part of it.
The wilderness, the dew drops, the pollen . . .
I became part of it.

(*Journal of American Folklore*, 1950)

I found no gorillas that day nor on the following one, although fairly fresh nests were common. On the third day I took with me N'sekanabo, the guard. He was a huge fellow with a ready grin and muscular arms that would do justice to a gorilla. Our paths led through dense stands of lobelias, each naked stem six feet tall and with a cluster of large leaves at the apex. When injured in some way the plants exude a sticky white fluid that tastes extremely bitter and, when inadvertently rubbed into the eyes, burns with blinding fury. Slightly ahead and to one side of us, we suddenly heard a quarrelsome, high-pitched scream, such as I have heard a captive infant gorilla give when deprived of a favorite toy. Motioning N'sekanabo to wait, I crept ahead and from the cover of a tree trunk looked out over a shallow valley. A female gorilla emerged from the vegetation and slowly ascended a stump, a stalk of wild celery casually hanging from the corner of her mouth like a cigar. She sat down and holding the stem in both hands bit off the tough outer bark, leaving only the juicy center which she ate. Another female ambled up with a small infant clinging to her back. She grabbed a stalk of wild celery near the base and pulled it up with a jerk. She then pushed a swath of vegetation down with one hand, squatted and ate, scattering the strips of celery bark all over her lap. Wild celery closely resembles the domestic variety, but, as I later found out, it tastes bitter. This plant turned out to be the second most important food item around Kabara. To obtain a better view of some of the other members of the group, I became incautious and let myself be seen by one female. She emitted one short scream and ran off into the undergrowth. A large youngster, weighing

about eighty pounds, climbed up the sloping trunk of a tree, looked in my direction intently, and descended rapidly. Suddenly seven animals, with a large silverbacked male bringing up the rear, filed by only a hundred feet from me. He paused briefly, peering at me from the cover of a screen of herbs, only the top of his head showing. After a harsh staccato of grunts, which apparently functioned as a warning to me as well as to the group, he hurried away, closely followed by three females and four youngsters, two of which rode piggy-back, clutching the females.

Kay and I saw the gorillas only from a distance on the following day, but on the third day I managed to creep to the edge of the group. Crouched in the low crotch of a *Hagenia* tree and partially hidden by some branches, I obtained wonderful glimpses of the apes as they fed and rested among some tall shrubs, completely unaware of my nearness. The silverbacked male moved leisurely up to a *Vernonia*, a tree-like shrub, grasped the stem about six feet above ground, and with a sudden jerk uprooted the whole tree. With the same slow motion he took a branch, splintered and tore it apart with his teeth, and gnawed the tender white pith with his incisors as if eating corn-on-the-cob. A female reclined on her back on a sloping tree trunk. She had pulled the crown of a nearby *Vernonia* close to her face and was eating the purple blossoms, plucking each one between thumb and index finger before popping it into her mouth. A large youngster climbed into the crown of a shrub and squatted there, swaying gently back and forth. Then it swung its legs free and hung there by one hand, rotating slowly.

A fearful scream shattered the tranquil forest. All feeding ceased as the animals raced to a thicket from which the sound came. Then others screamed with high piercing notes and milled excitedly about. Finally the silverbacked male emitted a series of deep guttural grunts, slow and emphatic. The screaming ceased. Later, after the gorillas had moved over a ridge and out of sight into another valley, I examined the area and discovered the reason for the commotion. A gorilla had apparently fallen some five feet into a cave. Footprints left when it scrambled to get out were clearly visible in the soft earth. A few days later I entered this lava cave with a flashlight and found that equally curious visitors had preceded me. At least three gorillas had entered the cave through a large opening obscured by hanging vines and had explored some hundred feet into the dark and clammy interior. A side chamber in this grotto contained a small lake, terribly still except for a constant drip, drip that seemed to come out of nowhere.

Group I, as I referred to the first gorillas I had seen in this area, was not the only group around Bishitsi at that time (see Table 1, p. 111). On August 22, I heard a male beat his chest about one hundred yards from the animals I was watching, and the following morning inspection of the forest revealed that no less than three groups had nested close to each other. One of these groups was large, comprising nineteen animals as a later count showed, but the third group contained only five gorillas — a male, two females, and two youngsters. These three groups, (I, II, and III,) remained in the same part of the forest for five days. Once groups II and III joined briefly, nesting together for a night, and twice groups I and II slept only fifty yards apart. But I was not yet experienced enough to track all the groups through the maze of criss-crossing trails, and the details of their interactions remained obscure to me.

Slowly, as I watched these groups during the latter part of August, I developed techniques for studying the gorilla. Every species of animal presents its own special problems, and not until these have been solved is an intimate study possible. Most of my visits to a group began at the site where I had last seen it the previous day. Cautiously I followed the trails of the gorillas through the trampled vegetation, never certain if the animals had gone a hundred yards, a mile, or doubled back so that now they were somewhere behind me. The trails always had an interesting story to tell, and in many ways I enjoyed my saunterings as much as the sight of the apes themselves. Almost unconsciously I fell

TABLE 1. THE COMPOSITION OF GORILLA GROUPS AT KABARA*

GROUP	Silverbacked Male	Blackbacked Male	Female	Juvenile	Infant	TOTAL
I	1	0	3	2	2	8
II	1	3	6	5	4	19
III	1	0	2	1	1	5
IV	4	1	10	3	6	24
V	2	2	3	2	2	11
VI	1	1	9	2	7	20
VII	1	2	6	4	5	18
VIII	1	2	8	3	7	21
IX	4	3	9	5	6	27
XI	1	1	6	2	6	16
Total	17	15	62	29	46	169
Per cent	10.0	8.9	36.7	17.2	27.2	

* Compositions represent animals present at the time of the first complete count of each group.

into the same unhurried pattern of movement as that of the gorillas, especially when the morning sun shimmered on the leaves and the mountains reached serenely into the sky. When gorillas were feeding, they fanned out, leaving many trails littered with discarded celery bark and other food remnants. When the gorillas were traveling, they moved in single or double file, only to rest after awhile close together on an open slope. Sometimes a musty odor, like that of a barnyard, permeated the air, and I knew that it was the site where the animals had slept the night before. The nests and the area around them were littered with dung, and hundreds of small brown flies darted around laying tiny white eggs on the moist surfaces of the feces. It often took me over half an hour to find all the nests, since a gorilla occasionally slept sixty feet or more from its nearest neighbor. I mapped each site and paced off the distance between nests. By measuring the diameter of dung in the nests, I could determine the resting place of large silverbacked males and of juveniles. Frequently both medium and small-sized sections of dung lay side by side in the same nest, indicating that a female and her offspring had slept in it.

Somewhere not very far ahead in the undergrowth were the gorillas, often without a sound to reveal their presence. To track them over this last piece of trail, never certain when a shaggy head would rear above the vegetation, never certain that an attack by the male would not follow careless approach, was the most tense and exciting part of the day. Cautiously, I took one, two, three, steps, before stopping, all senses alert, listening intently for the snapping of a branch or the rumbling of a stomach. I climbed up fallen logs or into the low branches of trees to scan the forest ahead for a glimpse of a black body among the weeds. For minutes I stood motionless, nerves so keyed to receive the slightest stimulus that even the distant whirr of a sunbird's wing was enough to startle me.

Frequently my first intimation that the gorillas were near was the sudden swaying of a lobelia or a branch, jarred by a passing animal. Then there were two courses open to me: I could hide myself and watch the gorillas without their being aware of my presence, or I could remain in the open with the hope that, over the days and weeks, the animals would become accustomed to seeing me near them. The former method had a great advantage, since the behavior of the apes was not influenced in any way by my intrusion. But I soon found out that I often lost useful observations when I hid myself too well, for if I tried for a better view, the animals saw me and grew excited. I usually walked slowly and in full view toward the gorillas and climbed up on a stump or tree branch where I settled myself as comfortably as possible without paying obvious attention to the animals. By choosing a prominent observation post, I was not only able to see the gorillas in the undergrowth, but they could inspect me clearly and keep an eye on me.

Animals are better observers and far more accurate interpreters of gestures than man. I felt certain that if I moved around calmly and alone near the gorillas, obviously without dangerous intent toward them, they would soon realize that I was harmless. It is really not easy for man to shed all his arrogance and aggressiveness before an animal, to approach it in utter humility with the knowledge of being in many ways inferior. Casual actions are often sufficient to alert the gorillas and to make them uneasy. For example, I believe that even the possession of a firearm is sufficient to imbue one's behavior with a certain unconscious aggressiveness, a feeling of being superior, which an animal can detect. When meeting a gorilla face to face, I reasoned, an attack would be more likely if I carried a gun than if I simply showed my apprehension and uncertainty. Among some creatures — the dog, rhesus monkey, gorilla, and man — a direct unwavering stare is a form of threat. Even while watching gorillas from a distance I had to be careful not to look at them too long without averting my head, for they became uneasy under my steady scrutiny. Similarly they considered the unblinking stare of binoculars and cameras as a threat, and I had to use these instruments sparingly. As could be expected, gorillas were more annoyed and excited on seeing two persons than one. Kay had to remain home much of the time, and the park guard, who sometimes accompanied me, had to hide himself while I watched the apes. I decided not to follow the animals once they had moved from my sight, for pursuit could easily frighten them and increase the chance of attacks. In general, I put myself into the place of a gorilla and tried to imagine to what actions I would object if suddenly a strange and potentially dangerous creature approached me. In all the months I spent with the gorillas, none attacked me.

Establishing rapport with the gorillas was fairly easy, because their senses are comparable to those of man — not man of the city who is unable to react to the subtle stimuli of his surroundings amidst the incessant noise of the machine, but man attuned to the wilderness. As with man, sight is the most important sense in gorillas. The apes are very quick at spotting slight movements, and often they watched my approach before I was even aware of their presence. Hearing, too, is well developed in gorillas, but they respond only to strange sounds, like the human voice, or sounds out of context. When a group fed noisily, my approach could be rather casual, but when they were resting all were alert to a stray branch snapping underfoot. The sense of smell seems to be relatively poor in gorillas: they rarely responded to my presence even when I was downwind and within fifty feet of them. Twice, however, gorillas seemed to smell me when I sweated profusely — but, as Kay commented, it does not take an acute sense of smell to do that.

On several occasions resting animals became uneasy while I was watching them from a distance. I was upwind, completely silent, and they obviously had not seen me. Yet they seemed to sense that something was not quite right. Perhaps they responded to subliminal stimuli, too vague to be assimilated consciously; or perhaps some other sense warned them of possible danger. I have had similar experiences when wandering through the forest. Suddenly I had the feeling, in fact I knew, that gorillas were close by, yet I had neither seen, heard, nor smelled them. More often than not, I was correct. Most naturalists, I feel sure, have had similar encounters with animals.

On the last day of August, as N'sekanabo and I clambered up the boulder-strewn depths of Kanyamagufa Canyon, a tremendous roar filled the chasm and bounding from wall to wall descended the mountain like the rumbling of an avalanche. We started and ducked and then peered up at the silverbacked male who, surrounded by his group, stood motionless at the canyon rim looking down at us. Quietly and as unobtrusively as possible we retraced our steps under the watchful eye of the male, feeling chastised like children for having so crassly intruded into his domain. We climbed up the opposite wall of the canyon, and later, when I was able to see the gorillas well, I found to my delight that they were old acquaintances. Six months previously, in March, Doc and I had watched this group for several hours. Almost daily throughout the month of September I visited these animals, group IV, watching them, enjoying their antics, and worrying over their problems. All the members in the group became definite individuals whom I recognized and named. No other group taught me as much or took a greater hold on my affection.

They moved up the precipitous slopes of Mt. Mikeno after our first meeting, going higher and higher until the *Hypericum* trees grew stunted and timber line was not far above. I ascended the mountain daily to be with the animals, my senses vibrant and alive as I clambered up. The silent forests were another world from the villages and fields that lay far below us. On the horizon, past Goma and Lake Kivu, dense ranks of clouds gathered, as they usually did by mid-morning, drawing closer and closer until their advance was halted by the ramparts of the mountains. But drawn inexorably onward they stormed soundlessly up the slopes, fingering the canyons, dodging from tree to tree, until finally they had gathered everything into their clammy embrace.

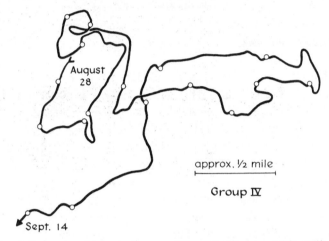

The daily route of travel of group IV on the slopes of Mt. Mikeno between August 28 and September 14, 1959. Each circle represents one night nest site.

On September 4, I came upon the gorillas feeding slowly on a steep slope about a hundred yards above me. I sat down at the base of a tree, and, with binoculars resting on my drawn-up knees, I scanned the slope, trying to pinpoint the whereabouts of the four silverbacked males in the group. The large male gorillas are the most alert, unpredictable, and excitable members of the group and hence the most dangerous. Squatting with his back toward me was Big Daddy, easily recognizable by the two bright silver spots on his gray back. As he turned to rest on his belly, he saw me, gave me an intent look, and emitted two sharp grunts. Several females and youngsters glanced from the vegetation in his direction and then ambled to his side, warned that possible danger was near. Big Daddy was the undisputed leader of the group, a benign dictator who by his actions determined the behavior of the other animals. He stood now looking down at me with slightly parted lips, his mighty arms propped on a knoll, completely certain of his status and his power, a picture of sublime dignity.

D. J. was the striving executive type who had not yet reached the top. He was second in command, a rather frustrating position from a human point of view, for in such matters as determining the direction of travel and the time and duration of rest periods the females and youngsters ignored him completely. He lay by himself on his back, one arm slung casually across has face, oblivious to the world.

The Outsider roamed slowly around the periphery of the group, intent on his own doings. He was a gigantic male in the prime of life, visibly larger than Big Daddy, and by far the heaviest male around Kabara. His nostrils were set like two black coals in his face, and his expression conveyed an independence of spirit and a glowering temper. His gait was somewhat rolling like that of a seaman, and with each step his paunch swayed back and forth. To estimate accurately the weight of gorillas in the wild is difficult, but I believe that the Outsider must have weighed between four hundred and fifty and four hundred and eighty pounds. Gorilla males are often said to weigh six hundred pounds or more, but these are the weights of obese zoo animals. Two mountain gorillas in the San Diego Zoo, for example, weighed that much, and they gained still more before they died. In contrast, of ten adult male mountain gorillas killed and weighed in the wild by hunters and collectors, the heaviest animal reached four hundred and eighty-two pounds, and the average was about three hundred and seventy-five pounds.

The fourth silverbacked male in the group was Splitnose, so named for the ragged cut that divided the upper part of his left nostril. He was young, his back barely turned silver, and he lacked the quiet reserve, the sureness of action, which characterized the other three adult males. As if to compensate for his uncertainty of mind, he was highly vociferous whenever he saw me, roaring again and again, sounding his warning over the mountains. But none of the other animals responded visibly.

Mr. Dillon (left), *the leader of group* VI, *sits and yawns, showing his black, tartar-covered teeth. Ten animals are visible in the dense under-growth.*

Two Bantu women and their children rest near a banana grove. The woman on the right is having her head shaved by the girl.

An old secondary forest near Utu in an area last inhabited by man over forty years ago.

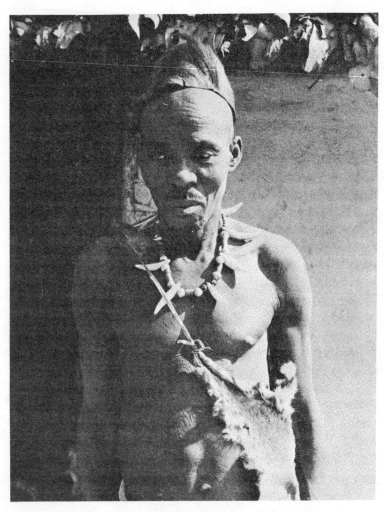

A medicine man in one of the villages in the Maniema Forest. His
necklace consists of beads and leopard teeth.

Apparently D. J. had hatched a plan, for suddenly he left his resting place and circled uphill. Then stealthily, very stealthily, he angled toward me, keeping behind a screen of shrubs. But gorillas are not very good at this sort of thing. Branches broke underfoot and to orient himself he stood up to glance over the vegetation. As soon as I looked directly at him, he ducked and sat quietly before continuing his stalk. He advanced to within thirty feet of me before emitting a terrific roar and beating his chest. Immediately afterward, before the echo of the sound had died away, he peered out from between the bushes as if to see how I had responded to his commotion. Never, even when I fully expected it, was I able to get used to the roar of a silverbacked male. The suddenness of the sound, the shattering volume, invariably made me want to run. But I derived immense satisfaction from noting that the other gorillas in the group startled to a roar just as visibly as I did.

With the male only thirty feet from me, I became uneasy and thought it prudent to retreat to a safer place. Cautiously, I ascended a tree to a height of ten feet. One of the ten females in the group left Big Daddy and ambled to within seventy feet of me to sit on a stump, her chin propped on her folded arms. Slowly, as if daring each other to come closer, the whole group advanced toward my tree. I felt a brief spasm of panic, for the gorillas had never behaved in this manner before. They congregated behind some bushes, and three females carrying infants and two juveniles ascended a tree and tried to obtain a better view of me through the interlacing vines that festooned the branches. In the ensuing minutes we played a game of peekaboo: whenever I craned my neck in order to see the gorillas more clearly, they ducked their heads, only to pop forth again as soon as I looked away. One juvenile, perhaps four years old, climbed into a small tree adjacent to mine, and there we sat, fifteen feet apart, each somewhat nervously glancing at the other, both of us curious, but refraining from staring directly to eliminate all intimation of threat.

Junior, the only blackbacked male in the group, stepped out from behind the shrubbery and advanced to within ten feet of the base of my tree, biting off and eating a tender leaf of a blackberry bush on the way. He stood on all fours and looked up at me, mouth slightly open. In all my hours with group IV, I was never able to fathom Junior completely. He was less than eight years old, still the size of a female, but his body had already taken on the angular and muscular build of a male. There was recklessness in his face and a natural mischievousness, which even his inherent reserve could not hide. At the same time his look conveyed a critical aloofness as if he were taking my measure and was not quite sure if I could be fully trusted. He was the only gorilla who seemed to derive any sort of satisfaction from being near me. Later in the month, rarely a day went by when he did not leave the group to sit by me, either quietly watching my every action or sleeping with his back toward me. Today he was still somewhat uncertain of himself as his indrawn and compressed lips showed. Man too bites his lips when nervous. Occasionally he slapped the ground with a wild overhand swipe, using the palm of his hand, then slyly looked up at me, apparently with the hope that his wanton gesture had been startling. The other members of the group rested quietly. Every fifteen or twenty minutes one of the males jerked out of his slumber to roar once or twice before reclining again to continue his nap.

All apprehension of the gorillas had long since left me. Not once had their actions portrayed ferocity or even outright anger. The silverbacked males were somewhat annoyed, to be sure, and several animals were excited, but all this was offset by their curiosity concerning me and their rapid acceptance of me. As long as I remained quiet, they felt so safe that they continued their daily routine even to the extent of taking their naps beside the tree in which I was sitting. Early in the study I had noted that the gorillas tend to have an extremely placid nature which is not easily aroused to excitement. They give the impression of being stoic and reserved, of being introverted. Their expression is usually one of repose, even in situations which to me would have been disturbing. All their emotions are in their eyes, which are a soft, dark brown. The eyes have a language of their own, being subtle and silent mirrors of the mind, revealing constantly changing patterns of emotion that in no other visible way affect the expression of the animal. I could see hesitation and uneasiness, curiosity and boldness and annoyance. Sometimes, when I met a gorilla face to face, the expression in its eyes more than anything else told me of his feelings and helped me decide my course of action.

The brief morning spell of sunshine had given way to dank clouds that descended to the level of the trees. For five hours I perched on a branch, chilled through and through, my fingers so stiff I was barely able to take notes. It began to rain heavily, and soon the rain turned to hail. The gorillas sat in a hunched position, letting the marble-sized stones bound off their backs. They looked thoroughly miserable with the water dripping off their brow ridges, and the long hairs on their arms were a sodden mass. I sat huddled next to the trunk of the tree, hoping for some protection from the canopy. My face was close to the bark, and I smelled the fungus-like odor of lichen and moldy moss. I could not leave the gorillas without disturbing them, and I had to wait until they moved away.

After the hail ceased and the rain was a mere drizzle, the gorillas spread out to forage. From behind a bush came a curious staccato sound, one which I had not heard before: a rapid series of loud *ö—ö-ö-ö*, with the first vowel forceful, emphatic, and separated from the others by a distinct pause. This sound was emitted over and over again, and after two or three minutes I became aware of the situation that elicited it. D. J. and a female were together. They were copulating. She rested on her knees, belly, and elbows, and D. J. was mounted behind, holding onto her hips. The male pushed, and since the slope was steep, the two animals moved downhill. They covered forty feet in fifteen minutes, with the female using her hands to part vegetation as they progressed. They stopped three times. D. J. was thrusting rapidly.

His vocalizations grew harsher, and the female screamed piercingly. The male now clasped her by the armpits, and he was nearly covering her back. They came to rest against a tree trunk, and a hoarse, trembling sound, almost a roar, escaped from D. J.'s parted lips, interrupted by sharp intakes of breath. He sat back, the act completed. The female lay motionless for ten seconds, then walked slowly uphill, while the male remained, panting. In spite of these far from silent doings, none of the other members of the group paid the slightest attention. Even Big Daddy, the boss, who rested in full view of the copulators, was seemingly oblivious of the spectacle.

At last the gorillas moved away and, after six hours on the branch, I was able to descend, stiff and cold. In spite of the inclement weather, I was elated by the perfection of the day.

During ensuing days the group traveled through several deep ravines parallel to the slope of Mt. Mikeno in the direction of Kabara. The gorillas had become used to my presence: the females hardly responded at all, and Big Daddy did little more than grunt briefly in annoyance each time he first saw me. Even Splitnose, though still his old vociferous self, was somewhat subdued. Junior seemed to await my arrival. A juvenile, perhaps four years old, frequently followed him closely, aping his every action. J. J., the juvenile, seemed to be male, although the sex of young gorillas in the wild is impossible to determine with certainty. He was a cute little fellow with a stomach taut as a drum, long unkempt hair on the crown of his head, and with an expression of sauciness that always seemed to herald some prank.

The guard N'sekenabo accompanied me one morning in search of the group. When I caught sight of the animals about fifty yards ahead, I motioned him to duck behind a tree while I swung up into a low branch. As usual, Junior sauntered over and twenty feet from me climbed the trunk of a fallen tree. He then moved along the log on all fours with curious abrupt and jerky steps. His body was stiff and erect, and his arms were curved outward at the elbows, giving them a bent appearance. Junior was strutting, seemingly trying to show off by making himself appear big and powerful. I was amused by his doings and had forgotten all about the guard. When I looked down, I saw N'sekenabo standing below clutching his machete with both hands, beads of sweat covering his forehead. His lips were gray with fear, but I admired his fortitude in remaining with me in the presence of the beast. Junior retreated casually, looking over his shoulder before hurrying after his retreating group.

On several occasions Junior, or sometimes another gorilla, approached within sixty feet or less and shook his head from side to side. It was an odd gesture, one that seemed to signify "I mean no harm." To see what gorillas would do if I shook my head at them, I waited until Junior was thirty feet away and paying close attention to me as I rewound the film in my camera. When I began to shake my head, he immediately averted his face, perhaps thinking that I had mistaken his steady gaze for threat. Then, when I in turn stared at him, he shook his head. We continued this for ten minutes. Once he relaxed the muscles of his lower jaw so that the shaking produced a rattling sound. Later, when I inadvertently met gorillas at close range, I employed head shaking as a means of reassuring them and they seemed to understand my good intentions.

On September 23 I witnessed another copulation. The animals were scattered over the slope, resting and snacking and absorbing the warm rays of the sun. The Outsider stood on all fours, seemingly scanning the valley below, when a female appeared out of the undergrowth behind him. She clasped him around the waist and thrust herself against him about twenty times. At first the male appeared oblivious but then he swiveled around, grabbed the female by the waist, pulled her into his lap, and began to thrust. Big Daddy, who was lying fifteen feet away, rose and slowly approached the pair. The Outsider desisted and retreated ten feet uphill. Big Daddy and the female then sat side by side, but when he ambled away after about one minute the Outsider returned to her. She gazed into his eyes, and there must have been something in that look, for he did not tarry. With the female in his lap he thrust rapidly, about twice each second, and soon emitted the peculiar call which I had heard during the previous copulation. The female twisted sideways and squatted beside the male, who rolled onto his belly to rest in this position for ten minutes. Suddenly he sat up and pulled the female onto his lap again, but she broke away. Both then rested for over half an hour. My field notes describe the completion of the act: After a prolonged rest, the female rises and stands by the rump of the male. He glances up, and they stare at each other. The process of pulling her into a sitting position and thrusting is repeated. At about seventy-five thrusts he begins his copulatory sound. His eyes are closed, and the thrusts rock her back and forth, a motion aided by his hands on her hips and the swaying of her body. His lips are pursed, and her mouth is slightly parted. At about one hundred and twenty thrusts the male suddenly opens his mouth with a loud, sighing "ahh"; the female opens her mouth at the same time. He relaxes, she rises and leaves.

I never observed another copulation in the wilds, nor did I see free-living gorillas copulate face to face, a position they use regularly in the Columbus, Ohio, zoo. Females apparently become sexually receptive only for three to four days during the time of ovulation, which occurs once a month. At such times they appear to initiate sexual contact with a silverbacked male of their choice, and this male need not be the leader and dominant animal in the group. Females are usually not receptive when in advanced stages of pregnancy and when lactating. Since most females are either pregnant or lactating, the silverbacked male or males in the group may on occasion spend as much as a year without sexual intercourse, for they seem to make no overtures to the females unless these indicate their receptivity.

Some scientists have maintained that monkey and ape groups remain together over long periods of time because the males have continuous and ready access to receptive females. But from my observation it seems that gorilla groups remain stable, on the whole, even though there may be no receptive females for months at a time, indicating that sex is of little or no importance here. Gorillas always gave me the impression that they stay together because they like and know one another. The magnanimity with which Big Daddy shared his females with other males, even though some were only temporary visitors, helped to promote peace in the group. Eskimos and some other native peoples also found that it caused less dissension to share their wives with visitors than to have them taken by force.

One of my problems was to determine the approximate age of each gorilla so that I might have some ideas about how long infants remain with their mothers and at what age the apes begin to breed. Fortunately, I had visited gorillas in several zoos in the United States and in Europe before coming to Africa. By comparing the weight and size of captive gorillas of approximately known age to that of free-living ones, I derived a very crude aging scale. At one year of age a youngster weighs fifteen to twenty pounds, at two years about thirty-five pounds, at three years about sixty pounds, at four years about eighty pounds, and at five years about one hundred and twenty pounds. Older animals show a great disparity in weight, due both to sexual and individual differences. Doc and I had seen an infant on the day of its birth in March, 1959, and I was able to watch its development until August, 1960, a period of seventeen months, and I also noted the growth of others for from ten to twelve consecutive months. All this information gave me a fairly accurate means of estimating the age of gorillas from birth to three years. During this period of their life the youngsters cling to or remain close to their mothers, and I classified them as infants. Between the ages of three and six years, gorillas are largely or completely independent of their mothers, traveling under their own power with the group, and I called them juveniles.

Information from zoos also revealed that males and females remain about the same size until they are from eight to nine years old. At this age, the males show a spurt of growth, which continues for some two to three years, with the result that fully grown adult males weigh roughly twice as much as females. Between the ages of nine and ten the males begin to show silver or gray hairs on their back. Therefore, blackbacked males are about six to ten years old, and silverbacked males are over ten years old. Males like Big Daddy, the leaders of groups, are probably at least twelve years old. The age of females can not be told with any certainty, and any female of six years or over I simply considered an adult.

One morning I found the trail of a single gorilla leading away from group IV. An hour later, along the precipitous walls of Kanyamagufa Canyon, I caught up with the Outsider, who roared and crashed away through the underbrush. He slept by himself for two nights, building his solitary nest of branches on the grounds. Then to my surprise he cut across the fresh trail of his former group, apparently followed it, and rejoined the animals after living alone for two days. Lone males leading a solitary life in the forest have been encountered by many visitors in gorilla country, and the popular supposition has always been that such males are poor, defeated creatures driven from groups by younger stronger rivals or, as expressed by the anthropologist Coon in 1962, "the youths are driven out of the family band at about the time of puberty, not so much because they could not feed themselves earlier, but because at that time they begin to arouse jealousy, in the well-known Oedipus fashion, in their parents." The behavior of the Outsider gave me my first intimation that these ideas, like so many other ideas about gorillas, might be quite wrong.

On September 9, I was again following the trail of a lone male in the vicinity of group IV when I met the Outsider coming down another path. He wheeled around and lumbered off, running slightly sideways with his head turned, thereby protecting his face from the undergrowth. Apparently the wanderlust again had taken hold of his independent spirit. The other trail led into a ravine, but I did not follow it. Later in the morning and farther up along the edge of the ravine, as I was quietly watching group IV, the head of a strange silverbacked male popped suddenly from the vegetation within thirty feet of me. He saw me, and his mouth dropped open in utter astonishment before he collected himself and with a wild roar dove over the rim of the canyon. Ten minutes later, apparently having gathered his courage, he walked past me with a determined air and joined the group. He sat down near Big Daddy, D. J., and Splitnose, who were resting together, without eliciting the slightest response from them. The Newcomer,

as I called him, was a young fellow, somewhat smaller than Splitnose. The group was apparently to his liking, for he remained a
member for at least eleven months. The complete indifference
with which the Newcomer had been accepted by the group was
striking, and I thought it probable that he had been a member of
this band before temporarily taking up a lone life. He confirmed
my earlier observation on the Outsider by showing that, in this
group at least, males could come and go as they pleased.

I had so far watched gorillas at various times of the day, while
they were resting and moving, playing and feeding, yet I had
never seen them as they bedded down for the night or as they rose
from their nests in the morning. The sky looked dark and ominous
when I left Kabara one afternoon with a pack on my back, prepared to sleep with group IV, which was at that time criss-crossing
a rather open slope high up on Mt. Mikeno. The group had
crossed one of the chasms that radiated from the mountain. The
walls were deeply carved and slippery, and only on some of the
rocky shelves had clumps of brush found a tenuous foothold.
Until now the gorillas had never ventured on a route likely to
deter me, but as I stood pressed against the moist stone wall,
grasping a fragile stem while angling with my foot for support in
a rocky niche, I admitted to myself that the apes were certainly
more agile than I. But even the gorillas slipped and fell on occasion. Once I followed the track of an animal along another ravine
and found that the footprints ceased at a bare sloping rock which
once had been covered with moss. Apparently the gorilla had
stepped on the rock, the moss had given way, and the gorilla
had tumbled some twenty feet, wildly grasping at the canyon wall.
The fall apparently caused no serious harm, for the tracks led
away from the base of the bluff.

Crossing the canyon had delayed me, and it was nearing five
before the crackling of the undergrowth revealed the presence of
the gorillas about forty yards ahead. Most of them sat silently,
looking subdued and somewhat sleepy, but others still snacked in
desultory fashion. I hid behind the spreading bole of a tree, anx-
ious to observe the gorillas without making them aware of my
presence. Four infants, between one and two years of age, played

exuberantly on a sloping tree, running up and down in single file in a game of follow-the-leader and sliding down the moss-covered trunk on their seats or on their bellies. A paunchy female walked up and, leaning against the trunk with folded arms, looked silently at the youngsters. Her look apparently conveyed "time to go to bed," for dutifully the youngsters stopped their game. One infant descended the trunk by pushing its way under the female's arms and out between her legs; another climbed on her head and cantered down her back before sliding feet-first down her rump; and a third took a running jump and with flailing arms and legs landed on his bottom on the female's back. Only the fourth infant sedately followed at her heels and went off to bed.

Big Daddy sat hunched by a shrub, motionless, like some unearthly being carved out of granite in the semblance of a man. He reached out with his right hand and bent a branch, which he pushed down under his left foot. Then for five minutes he slowly bent all branches and shrubs within reach and pressed them down without pattern or sequence, rotating slowly until he had constructed the rim of a nest around his body. After that he reclined on his belly with arms and legs tucked under, presenting his massive back to the drizzling rain that had begun to fall. As soon as Big Daddy began to build a nest, several other members of the group did the same. While a female was busy preparing a bed for the night, her infant, about two years old, climbed up seven feet into the crotch of a shrub. The youngster grabbed a branch with one hand and pulled it inward until it snapped. After pushing the branch down into the crotch and stepping on it, it broke in several others, all of which were pressed on top of each other until after ten minutes of labor a crude platform was completed. The infant then sat on its nest for about ten minutes and looked around before descending to cuddle close to its mother for the night. Youngsters rarely sleep by themselves until they are almost three years old, but they may build practice nests as early as fifteen months.

At half past five all movement among the gorillas ceased, and the black forms of the sleeping animals fused with their shadowy surroundings. The turf at the base of the tree behind which I stood was dry and protected from the rain by the slant of the trunk. Carefully I smoothed over the ground and placed on it my tarp and on top of that my sleeping bag. Finally, undressed and in the warmth of my bag, I ate a can of sardines and some crackers. Then I lay and listened to the soft rain pattering on the leaves, and I looked up at the night clouds moving by and at the beardlike lichen swaying gently from the trees.

It was five fifteen, and the mountains lay dark and silent. There was a faint line of light on the horizon. I climbed up into the tree and huddled on a branch, shivering in the morning coolness. After about fifteen minutes it was light. Still the gorillas slept, lying in their nests either on their sides or on their bellies. Not until seven o'clock, after the morning sun had climbed over the distant ridge, did two females leave their nests and slowly wander about. Apparently Big Daddy was awakened by the movement, for he sat up, blinked, and looked around before starting to feed. Soon all foraged actively, grunting and grumbling and belching with contentment. D. J. ascended a tree to the height of ten feet and sat there dozing. After awhile he took hold of a branch with his hands and swung his legs free. But the branch, unable to bear his weight, broke slowly to dump him on his rump in the vegetation below. The Outsider, who was wandering near the edge of the group, spotted me and emitted a sharp grunt, which two other males immediately answered. The gorillas resumed their feeding and moved out of sight at a leisurely pace.

The gorillas generally bedded down for the night at dusk and began to stir in the morning during the hour after sunrise, having slept some thirteen hours. They were silent at night, except for the rumbling of stomachs or breaking wind; I never heard them snore. When excited in some way, a male sometimes beat his chest during the night. Late one afternoon I came upon group IV after several animals had already gone to bed. With darkness I crept closer and crawled beneath a bower of vines at the base of a log, trying to find some protection from the rain that was falling as usual. From far uphill came the faint sound of a chest beat, given perhaps by a lone male. The Outsider responded similarly, and from then on throughout the night at varying intervals the *pok-pok-pok* echoed over the slopes. I was uncomfortable that night, for the ground was damp and hard. Once I was startled out of a dream in which a herd of buffalo came trampling down on my prostrate form, and once I thought I heard a rumbling stomach close by. A little after 5:00 in the morning I rose stiffly. As I waited for daylight, I became aware of a black form in the vegetation about forty feet away — a female gorilla in deep sleep. Silently I retreated to safer ground.

There is something immensely satisfying in living in one's own private domain, hidden in the vastness of the mountains, protected from the problems and tribulations of the outside world. Here one is free in body and spirit; here the web of experience is entirely of one's own weaving. I had work to do, and I did it to the best of my ability. Each morning I set off over the mountains, my mind receptive to the signs of the forest, seeking the apes which had become a part of my life. But my work never became a plodding routine, for each new day brought with it some new experience.

Group IV moved down into the zone of bamboo during the latter part of September, where in the maze of stems I was unable to observe them well. When a few days later they crossed Kanya-magufa Canyon to reappear in their old haunts on the slopes of Mt. Mikeno, one of the females carried a newborn infant. Carefully she held the helpless creature to her chest, supporting it at all times with one arm even when walking. Newborn gorillas are tiny, weighing only four or five pounds, and they are so weak that they are unable to hold on to the hair of their mothers for more than a few seconds. Their movements appear disoriented, and they have a vacant look, just like a human infant. At the age of one month, young gorillas begin to follow with their eyes the movements of other members of the group. Goma, a gorilla infant born at the zoo in Basel, Switzerland, first responded to known persons at the age of two months, and at about the same age Jambo, born at the same zoo, chuckled and laughed when tickled by the keeper. By the age of two and a half months, infants show a marked increase in movement. Their upper and lower middle incisors have appeared, and the youngsters now reach for and chew on branches and vines. For the first time the mother may place her offspring on the ground by her side and watch over it carefully as it shakily tries to crawl. In general, the rate of development of a gorilla infant is roughly twice as fast as that of a human baby, but considerably slower than that of monkeys, some of which can walk within a few days after birth.

I had been unaware that the female was pregnant, for her belly, like that of all gorillas, was so distended from the vast bulk of greenery she consumed daily that no unusual swelling was evident. At the Columbus, Ohio, zoo the ankles of a female were swollen after about seven months of pregnancy, but this again is not easy to detect in free-living gorillas. In fact, some aspects of the gorilla's sexual life, such as age at sexual maturity, cannot be readily observed in the wild. Several gorillas have fortunately grown to maturity in captivity, and three pairs have had offspring. Once it was thought that, because of the gorilla's large size, sexual maturity in females was not reached until they were ten to fourteen years old. But Christina, at the Columbus zoo, conceived at the age of seven years. Gorillas have a monthly menstrual cycle, averaging about thirty-one days between periods. Zoo records show that the females begin to menstruate between the ages of six and seven years. Males seem to reach sexual maturity between nine and ten years, on the average, at a time when the hair on the back begins to turn from black to silver and the weight increases rapidly. But a male at the Washington zoo had viable sperm and impregnated the female when he was only about seven years old.

On one occasion I caught up with group IV as it poked around in the depth of a ravine. Mrs. September, as I named the female whose infant was born during that month, sat near a small patch of bare earth. She cradled her newborn, now a week old, in her arms and gazed at it tenderly. Mrs. Bad-eye ambled up within two feet and looked at the naked infant in the arms of its mother. Mrs. Bad-eye seemed old, with a sad and dejected look about her, and she lacked an infant of her own. Somehow she had injured her eye, blinding it, and now the lens was opaque and a swelling grew around it as the months passed. Gently she reached out as if to touch the newborn, but the mother slapped her hand away. Mrs. September then knelt down and scraped at the exposed earth with her incisors, leaving long grooves in the soil. She picked up small particles of loosened dirt between the tip of the thumb and the side of the index finger and ate them. Both D. J. and Junior joined her, and all three ate soil for several minutes, scraping and picking and scraping some more. The following day I collected some of the soil and later had it analyzed. The sample was very rich in potassium and sodium salts.

Group IV was not the only group that occupied my attention during September. In the middle of the month I had found an-

Two Kabara nest sites, one compact, the other split.
  A. Group IV
  B. Group VII
Squares represent nests of silverbacked males, circles nests of black-backed males or females, circle within a circle nests of females with infants, triangles nest of juveniles. The arrow indicates the direction of travel.

other group, group V, near the Bishitsi bluff. It contained two silverbacked males, two blackbacked males, three females, two juveniles, and two infants. A few days later two more silver-backed males and two juveniles joined this group, raising the size of the band to fifteen. I suspected that the newcomers were former members who had temporarily gone off together. The boss of group V was a tremendous male with such a withdrawn, phlegmatic disposition that he rarely bothered to respond to my presence. The shaggy crest on his crown was so large that it tipped jauntily to one side. The bony crest overlaid with dense connective tissue serves as an attachment for the muscles that move the lower jaw. Humans, with their relatively small jaw and bulging head, have sufficient area for the attachment of these muscles without the need for a crest. Female gorillas, much smaller than males, lack crests or have only a small one.

The Old One of group V was one of two silverbacked males in the Kabara area who looked aged. He had a hunched look, and his pelage was almost totally grey. Only the hairs on his arms retained their full black luster. The skin on his neck hung loosely like dewlaps, and his face was sunken and the eyes tired. I wondered how old he was. Only a few gorillas have lived longer than twenty years in captivity because of the poor conditions under which they are usually kept. The record is held by Bamboo of the Philadelphia zoo who died at the age of thirty-four and a half. Massa, at the same zoo, was about thirty-two years old in 1963, a pathetic beast, sitting alone in his cement-and-steel-bar-prison, plucking the hair from his body for want of something better to do. Wild animals rarely reach their maximum life span. I suspect that few gorillas grow older than thirty years, and the average may be only twenty years. Compared to man's average life expectancy of seventy years in the United States, the gorilla seems to have a relatively short life. But it must be remembered that in 50 B.C. the average life expectancy of man was only twenty-two years and in the year 1800 it was only about thirty-six years. Even today many people who do not receive extensive medical care have a life expectancy of only thirty years.

The third silverbacked male in group V was the Eskimo, so named for his large round face and the slightly Mongoloid slant to his eyes. He was the most beautiful gorilla I had ever seen. His blue-black pelage sparkled as if he curried himself daily, and the hair on his back was not gray but a velvety silver like the morning frost on our meadow. The Eskimo followed the Old One in rank, and the Young One, whose back had barely turned grey, stood at the bottom of the pecking order or dominance hierarchy. When, as in groups IV and V, more than one silverbacked male is present, there is always a definite rank order. For example, dominance is expressed in claiming the right of way along a narrow trail or in displacing from a sitting place an animal of lower rank. Such a rank order is common in many group-living vertebrates, like chickens, cows, and man. Contrary to popular belief, a dominance hierarchy does not cause strife and dissension but promotes peace within the group, for it relegates each member to a certain status and position: every animal knows exactly where it stands in relation to every other animal.

Silverbacked males dominate all the other members of the group, for size and strength seem to determine the position in the rank order to some extent. Similarly, females dominate juveniles, and juveniles dominate those infants which have strayed from their mothers. Once, at the beginning of a downpour, a juvenile sought shelter beneath the leaning bole of a tree. It sat huddled against the trunk, looking out at the sheets of water that plunged down around its dry haven. But when a female hurried toward the tree, the juvenile vacated its seat and fled into the rain. As soon as the female had taken over the dry spot, a silverbacked male emerged from the undergrowth. He sat down beside the female and with one hand pushed her gently but determinedly until she was out in the rain and he under cover. This hierarchy finds its close counterpart in the human family, with the father as the master and the children in inferior positions, presumably responding to the commands of their elders. Trouble arises when the father or mother vacillates in the assertion of dominance, letting the young do as they please one day and forbidding them the next. The children then become uncertain of their status and insecure in their position. Gorilla males, although gentle and tolerant, take no back talk, and disputes are rare. Once a silverbacked male squatted on a log with a juvenile beside him. The male reached over and pushed the juvenile lightly with his forearm until the youngster moved about a foot. Five minutes later, the male rose and faced the juvenile, who ignored the male even when pushed lightly. Suddenly the male shoved the juvenile roughly, and it half-fell, half-climbed down the log.

Unlike silverbacked males in a group, females seem to lack a definite and stable rank order among themselves. It is perhaps significant that quarreling erupts mainly among the females, with silverbacked males taking no active part. I think, although I cannot be sure, that females have a changing rank order in which mothers with newborn or very small infants are dominant over mothers with larger infants. Then, of course, individual temperament plays a large role in dominance, for irascible members of the group are generally avoided.

Group V was unusual among the ten groups I ultimately came to know around Kabara in that the males outnumbered the females. In the Kabara groups as a whole the ratio of females to silverbacked and blackbacked males combined was roughly two to one, a state of affairs also found in several other primate societies. In group V the situation was reversed. Among the females was Callosity Jane, so known for the ischial callosities or sitting pads on her rump — common in monkeys and rare in gorillas. Mrs. Greyhead looked old. The hair on her head was grey, and her shoulders were peppered with white. But she carried a small infant about four months old, and this she handled with the efficiency of a practiced mother. The Scabby One had a skin rash on her neck which she aggravated by scratching; her body hair lacked luster and was sparse. Although she had no infant when I first met her and still lacked one a year later, a juvenile about three and a half years old stayed near her. In its pinched cheeks and the shape of its nostrils this juvenile had a certain resemblance to the Scabby One, and I felt confident that she was its mother.

It has frequently been asserted that one difference in the social behavior between man and the apes is that young apes remain with their mothers only until they are weaned and can fend for themselves. In man, on the other hand, the bond between mother and infant is prolonged. Gorilla babies are weaned by the age of eight months, although they may go back to their mother for a brief suckle until they are eighteen months old. By the age of two or two and a half years, infants can travel under their own power fast enough to keep up with a moving group. Yet here was the Scabby One and her large youngster still side by side, obviously fond of each other. Sometimes she placed her arm over the shoulder of her offspring and cuddled it close. The age at which the bond between mother and young is severed varies. A female gives birth only once every three or four years unless, of course, her child dies in infancy. If the youngster survives, it remains with its mother about three years. But I saw one juvenile, about four years old, which continued to travel with its mother until she had a new infant. A close association may even persist after a new infant is born. One female in group VII, for instance, had an infant about eight months old. A juvenile frequently sat by her, and once she reached over and pulled it to her chest. Occasionally at night this juvenile crawled into the nest with its mother. Thus, a strong social tie between gorilla females and their young may persist for from four to four and one half years, or long after they have ceased to provide food, transport, and protection.

Late in August, as I was sauntering through the forest, the brush suddenly swayed fifty feet ahead and a number of gorillas, apparently disturbed during their noon siesta, ran a few steps on their hind legs, looking briefly back at me over their shoulders before racing silently away on all fours. The silverbacked male stepped behind a bower of vines and thus hidden in ambush waited for me. Fortunately, I was alert, and I responded to his tactics by standing motionless. The seconds passed, then the minutes, each of us waiting tensely for the other to make the first move. Without warning the gorilla rose to his full height, his gigantic arms reaching toward the sky, gave one shattering roar, and with the same movement melted into the undergrowth to follow his group. Such was my first meeting with group II. In the ensuing days the animals headed steadily away from Kabara, and I had to give up the group without becoming closely acquainted with it.

The group returned to its former haunts a little over a month later, and I spent many delightful hours in its company. There were nineteen gorillas including the silverbacked male, three blackbacked males, six females, five juveniles, and four infants. This group contained an old female with sagging, wrinkled breasts and with kindly, tolerant eyes that always seemed to carry a bemused and faraway expression. She lacked an infant of her own, but her affection found an outlet in Max. Max was an impish and rambunctious six-month-old infant who could never sit still. When his mother held him, he poked at her eyes with his fingers until she averted her head. He stiffened and arched his back, and, as soon as his mother loosened her hold, he twisted and wriggled until finally she placed him on the ground beside her. Then, as often as not, he stood by her side, hands raised above his head, wanting to be picked up again. When this gesture was ignored, he bumbled off under the watchful eye of his mother, determined to find entertainment. Sometimes a lobelia stalk became his temporary gymnasium, as he climbed up a few feet only to slide down with a whoosh, up again and down, over and over. If the old female was nearby, he hurried into her arms, and on one occasion he suckled on her flaccid breasts. One sunny morning, as the female slept on her belly, Max climbed up on her back, walked forward and stood on her head, slid off, and then, as she rolled over, climbed upon her abdomen. She grasped Max and held him against her belly with one hand, and he struggled and squirmed and tried to free himself. His mouth was partially open, with the corners pulled far back into a smile. The female loosened her hand, and Max grabbed it and gnawed at her fingers. She then toyed with him, touching him here and there as he attempted to catch her elusive hand. Finally Max lay on his back on her belly, waving his arms and legs with wild abandon, and the old female watched the uninhibited youngster with obvious enjoyment. Suddenly Max sat up and, with arms thrown over his head, dove backward into the weeds.

Moritz, a seven-month-old infant, arrived on the scene. He walked up to the old female and pulled her leg. Max rushed from the undergrowth, tumbled over Moritz, and yanked his hair. Continuing his charge, he raced up the belly of the female and jumped at her face. She sat up and gathered Max into her arms and rocked him gently back and forth, his small body looking very tiny and fragile and almost lost against the bulk of her body. Nearby, a dense curtain of vines formed a small cave at the base of a tree. Moritz hovered near the entrance of the cave as if guarding it. Max cantered up, prepared for battle. Moritz rose on his hind legs, beat his chest, and with arms held above his head, jumped at and fell over Max. They wrestled, arms and legs flying; they mock-bit each other's shoulders, pretending ferocity. Finally, Moritz threw a headlock on Max. Max pushed and pulled; he rolled over and managed to break free. Before Max could recover, Moritz dashed back to the cave. Max bullied his way past Moritz, and the two wrestled in the bower of vines. Max, apparently feeling himself the victor, stalked away, and Moritz rushed up from behind and rammed Max so hard with his shoulder that both sprawled. Immediately Moritz dashed back to the cave and emerged with a dry piece of wood under one arm. Max ran in pursuit, grabbed Moritz by the rump, and both wrestled again. A hulking juvenile walked through the entrance of the cave, tearing down the whole structure. On seeing their cave destroyed, Max and Moritz teamed up and swarmed all over the juvenile, who wrestled lightly with the two infants.

At the end of September, I assessed the results of my work in the month and a half we had spent at Kabara. I had found five gorilla groups and watched their behavior for over fifty hours, learning many new things about the life of this ape. The gorillas had proved amiable far beyond my wildest expectations, and in the coming months I hoped to become still more closely accepted by the various groups. Many of the unsolved problems, which at the beginning of the study had plagued me solved themselves. I had learned to track gorillas by their spoor, and I was now able to remain with a group day after day, tracing its precise route as it moved up the slopes and down into the valleys. The ability to recognize individual animals is essential in a detailed study of social behavior. To my delight, I soon noted that gorillas vary as much in their physiognomy as humans do. Once I had become accustomed to the black countenances, I had little trouble in recognizing old friends even after an absence of several months.

CHAPTER 7

A Home in the Highlands

Early in the morning, at six o'clock, a pair of white-necked ravens swooped down through the mists to land with a crash on the metal roof of our cabin, only to slide down the steep incline, scraping their nails along, to finally come to rest in the rain gutter, where they noisily wiped their bills. This morning visit effectively startled us awake every time. Kay fed the birds scraps, and within a few weeks they grabbed food from her hand and at last, to her delight, they landed on her arms when she whistled to them. Over the months the ravens grew as tame as our chickens, waddling with their rolling gait around our doorstep and even jumping on the table in our kitchen, always waiting for a hand-out.

$\mathbf{A}$t Kabara we found a freedom unattainable in more civilized surroundings, a life unhampered by a weight of possessions. We needed no keys to open locked doors or identification cards to obtain the things we desired. We, who had lived in America where the electric toothbrush and the battery-operated pepper grinder are becoming symbols of a civilization, moved backward in time to note with joy how little was needed for contentment. The beauty of the forest and the mountains laid claim to us, and as the days went slowly by we came to live for the moment, taking limitless pleasure in the small adventures that came our way. Our life was dictated by the vagaries of weather. We usually wakened with the first light of day and lay listening for the tinkling of rain on the tin roof and the moaning of the wind in the eaves. If it rained, we remained as we were, only our noses exposed to the coldness of the room. To hurriedly slip into our clammy clothes required courage, for the high humidity made the near-freezing temperature seem colder than it was. Usually I was up first, splashed water on my face, brushed my teeth, and shivering slightly opened the front door.

"Andrea," I called, and from the depth of the adjoining shed came a muffled "Bwana" in answer.

While Andrea built a fire in the stove, I walked into the meadow, my eyes invariably drawn to the summit of Mt. Karisimbi. Occasionally new snow covered the peak, giving it the appearance of a white sea shell and explaining why the Africans call it "mountain of the shell." Standing on a boulder near the hut, I scanned the meadow for duiker and buffalo and looked for tracks in the dew-soaked grass. Then I released our few chickens from the shed and threw them a handful of maize. We had brought the chickens with the hope of obtaining fresh eggs as well as meat, but after being transported to these high altitudes the hens laid one egg and never did so again. They lost weight, their flesh became stringy. Every week our flock grew smaller as one member after another was sacrificed to the pot. Meanwhile, Kay had started breakfast, and later, huddled close to the stove, we ate our bowl of hot oatmeal topped with raisins and brown sugar.

The best time to watch gorillas is after they have foraged for about two hours and are ready for their mid-morning siesta, between nine and ten o'clock. At that time they are satiated and little inclined to move away, and they can be observed for several hours. When the gorillas were far away, high on the slopes of Mt. Mikeno or over near the Rwanda border, I set out immediately after breakfast, leaving Kay to do the chores. Usually I wound my way along animal trails, the highways of the forest, where various creatures had left their tracks in the soft earth. Sometimes I found the ratlike prints of a mongoose or the dainty track of a genet cat. Occasionally I came across the fresh track of a leopard. Several leopards ranged through the Kabara area, ascending to altitudes of over thirteen thousand feet. I may have walked by them, or even under them as they lay crouched on a branch above the trail, but I never saw them. These magnificent cats have an evil reputation in gorilla country, for they supposedly stalk young gorillas and kill large numbers. Yet in our year at Kabara only one infant gorilla disappeared, and the cause of its death remained unknown. Diligently I examined every piece of leopard dung, looking for bones, black hair, or other sign to indicate that the cat had preyed upon gorillas. But I found only the remains of duiker and hyrax. In West Africa the hunter Fred Merfield watched a leopard jump to the ground in full view of a group of gorillas without alarming them. Baumgartel, in Kisoro, found not the slightest indication that leopards harassed his gorillas between 1956 and 1960.

And yet as soon as I had made up my mind that leopards probably do not kill gorillas, an exception promptly occurred. From Kisoro came a fascinating tale of a black leopard which had become a killer of gorillas, just as some lions prey on man. In February, 1961, the guide Reuben and his two trackers noted that some birds on the slopes of Mt. Muhavura were excited about something. On reaching the spot they heard noises such as leopards make from behind a bush, but they did not see the animal. Instead they found a duiker lying in its blood. And a little farther away was a dead silverbacked gorilla with severe wounds in the neck and with a gash in the right groin that laid bare the intestines. Following the trail of flattened vegetation uphill, Reuben found the spot where the male had apparently been attacked in his night nest by the leopard. Both rolled down the slope to the spot where the body was found. Three days after finding the male, Reuben discovered the decaying and partially eaten corpse of a female. The game department of Uganda attempted to hunt down the leopard, but the cat proved too elusive. Later in the year, in July, Dr. D. Zimmerman, of New Mexico Western College, watched a black leopard as it stalked a group of four gorillas high on the slope of Mt. Muhavura. The leopard crept to within three hundred feet of the apes before they moved off. Then the leopard vanished.

Occasionally, at long intervals, I discovered the doglike track of a hyena, a scavenger which rarely ascends to these heights. I was sorry for this since there were no other efficient scavengers in this region to rid the forest of dead animals. In September, 1959, some disease killed off many duikers, and almost daily I came across one or more bodies lying untouched in the bush. Also, I like hyenas: I like the blunt powerful head, the alert little eyes, and the soft gray and black coat; I admire their stamina and the way they have made the best of an ungainly body, successfully living on carrion, young antelopes, or any large animal they can pull down, and, if these items are not available, on boot leather and frogs; I also think much of their courage, for they have been known to drive lions from their kill and to snatch food from the camps of man, who in Africa shoots hyenas whenever he can, usually for no purpose or reason. In May we met three adult hyenas and three half-grown pups in a grassy area surrounded by brush in Queen Elizabeth National Park. As we watched, a female lay down and rolled over, and a pup came up and suckled on her engorged mammaries. When she rose, the young continued to nurse, and she lay down again, paying us scant attention. Another female rolled on her back and invited a pup to suckle. It was a peaceful family scene, one that I always remember when I see hyenas described as disgusting and cowardly. Because of their habits many scavengers have aroused the ire of man. Even naturalists have sometimes fallen into the habit of writing about vultures, wolverines, hyenas, and other scavengers in a derogatory fashion, with the result that the various ill-fitting terms have been taken as a factual appraisal of the animal's nature. It should be pointed out that the bald eagle, symbol of America's might, and the bear, appearing on many a coat of arms, are also scavengers on occasion.

The most abundant of the large mammals in these mountains was without question the red forest duiker. These antelopes commonly came to our meadow at dawn and dusk to feed on the green grass. Timidly they bounded away at our approach, and they also fled from the gorillas, scurrying silently away through the tunnels in the undergrowth. During the midday hours they usually rested in some hidden niche, in a bower of vines or at the base of trees where the ground was dry and soft. The duiker's life is a solitary one. I never saw more than two together. On October 1, on a steep slope near Kanyamagufa Canyon, I watched two adults courting. The male stood erect with neck stretched and nose pointing skyward. Ten feet away and facing him was the female. She bobbed up and down for twenty seconds. Suddenly the male bounded three feet into the air, his back arched like a bucking bronco. Again and again, about ten times, he gave these curious leaps. The female too bucked twice, then raced uphill, closely followed by the male. When she stopped, the male sniffed her tail; she swiveled around, faced him briefly, bucked once, and scampered away, to disappear in the bush.

Duikers have but a single young. For several days after its birth it remains concealed in the dense vegetation. On New Year's Day, as Kay and I were winding our way through the blackberry brambles and shrubby trees that grow along the edge of the Rukumi meadow, we found a tiny duiker, probably less than a week old. It could only walk shakily on its spindly legs, and when it grew alarmed it bleated like a sheep. On April 30, I heard the same sound far up the slope on Mt. Mikeno. After considerable searching, I spotted an infant duiker under some brush, soaked to the skin from a recent downpour and shivering violently. It was very young, probably less than three days old: its hooves were still soft, and a tattered remnant of the umbilical cord hung from its belly. Later, as they grew older, they accompanied their mothers, and especially during April and May I frequently saw half-grown duikers at the side of an adult.

Throughout our stay in Africa I was eager to observe the various kinds of monkeys and compare them with the gorilla. But only the golden monkey (*Cercophithecus mitis kandti*),

a close relative of the blue monkey, frequented the high altitudes, especially the bamboo zone, in the Virunga Volcanoes. These cat-sized monkeys with their golden orange bodies and black legs, crowns, and shoulders were among the most beautiful I had seen. In late March we watched for several days a group of thirty-three golden monkeys in the bamboo on the slopes of Mt. Muhavura. At dawn the males, females, and young began to forage in the tops of swaying bamboo stems, pulling the leaves close and taking five or ten rapid bites before sitting back and chewing. They were agile creatures, sometimes hanging briefly upside down by their hind legs alone to reach some tender bamboo leaves below. After the morning feeding was over, they groomed themselves and each other, and the youngsters jumped in play. When a hawk sailed over, one member of the group emitted a loud, humming, resonant *brum-brum-brum*, and the others chimed in until the slope resounded to their calls. But they did not seek cover. In the Mts. Visoke–Sabinio saddle area I came on a group of about sixty golden monkeys. When I approached, they broke into two units and fled in opposite directions. Most of the groups I saw were much smaller, generally from five to twenty animals each. Although these monkeys are primarily found in forests of bamboo, the leaves and small shoots of which constitute their main food, stragglers, lone animals or groups of two, occasionally move upward into the *Hagenia* woodland. Twelve times I saw or heard them around Kabara. Often they indicated their presence solely by a sharp *tshio*, a noise they made when disturbed; at other times I flushed them from the ground or low shrubs where they fed on *Galium* vines and the blossoms of *Vernonia*. On one occasion, Kay and I visited group VIII, far over near the Rwanda border. The gorillas were resting on a slope about forty yards away as we clambered up into the crotch of a tree, where Kay settled herself in an abandoned gorilla nest. A golden monkey fled through some shrubs about ten feet above the gorillas, and another called from a distance, but the gorillas completely ignored them.

Sometimes, after a morning with the gorillas, when the sun

shone and the mountains reared into a clear sky, Kay and I wandered in our meadow, looking, listening, or doing some small chores. Kay often washed clothes, scrubbing them by hand in a tub. I sat in the doorway and greased my boots, soaking in the sun, so rare at Kabara. Andrea and the guard spread their cloaks on the grass and lay down, smoking and talking, and the chickens foraged around us, clucking and scratching.

The bane of our life was the stove. It had no damper, and the flames roared up the pipe. The wood we burned was filled with coal tar, and the stove pipe became clogged with a black, brittle residue. Then the wood only smoldered, and the fire gave no heat. One of our routine jobs was to dismantle the stove and clean it.

To become familiar with the smaller creatures of the forest, those that scurried at night beneath the cover of herbs, I set a few traps. I found that seven kinds of small rodents and three species of shrews lived at the edge of the meadow, using common runways. By far the most abundant was a rusty-red vole with an orange belly (*Lophuromys aquilus*), which resembled a meadow mouse. Unlike other species, this vole was somewhat diurnal. It bred throughout the year, and it usually had only one or two young instead of a litter of four or more, as is common in mice of more northern climes. Fairly abundant too was a mouse with a greyish-brown back and a dirty yellow belly (*Lophuromys sikapusi*), and another mouse (*Praomys*), with lively black eyes and a black stripe running down the center of its tawny back. One of the rarer and most appealing of the rodents was a grey dormouse (*Claviglis*) with enormous eyes looking gargoyle-like from its rather squat face. By far the most beautiful of the mice was *Thamnomys*, with a sleek belly that was strikingly white and a back that was golden as honey. Beside it *Otomys* looked like a tramp with its brown hair unkempt and ragged. But *Otomys* had the distinction of being the largest of the voles around Kabara, ten inches long from its shiny black nose to the tip of its stubby tail.

Birds were not very conspicuous along the meadow's edge. There was the double-collared sunbird and a few other species, most of them small and shy. Occasionally a Ruwenzori turaco ran chattering like a squirrel along a branch, the bird's green plumage blending wonderfully well into the forest until it flashed its wings to reveal crimson undersides. Hawks (*Buteo oreophilus*) frequently soared high overhead, riding the updrafts. Olive thrushes, resembling the American robin in size and to some extent in the color of their reddish plumage, chattered and scolded in the shrubbery. On May 2, I found a nest of this species near the cabin in the fork of a *Hypericum* branch. The nest was made of sphagnum moss and lined with grass, and two olive-green and brown-splotched eggs nestled in the cup. Two weeks later, on May 16, two plump young birds, their eyes open, crouched in the nest when I peered over the edge, and by May 24 they were well feathered out and ready to leave the nest.

The most regular visitors to our meadow were a pair of white-necked ravens, lovely birds with iridescent black plumage and a striking white collar around the neck. The day after our arrival they came and sat in the tree by our cabin, watching our doings with evident interest. And almost daily thereafter they spent an hour or more with us. Early in the morning, at six o'clock, they often swooped down through the mists to land with a crash on the metal roof only to slide down the steep incline, scraping their nails along, to finally come to rest in the rain gutter, where they noisily wiped their bills. This morning visit effectively startled us awake every time. Kay fed the birds scraps, and within a few weeks they grabbed food from her hand and at last, to her delight, they landed on her arms when she whistled to them. Over the months the ravens grew as tame as our chickens, waddling with their rolling gait around our doorstep and even jumping on the table in our kitchen, always waiting for a hand-out. The right to every bit of food was victoriously disputed with the chickens, and once when a duiker wandered out on the meadow it too was forced to retreat. By our cabin, on a little rise, was a boulder, and on it the ravens liked to preen in the morning sun. Sometimes one or the other lowered its wings slightly, ruffled the feathers on its head, spread its tail, and emitted a bell-like, gurgling call of such softness and beauty that it was difficult to realize that the raven's usual voice is a hoarse croak. The female, who was indistinguishable from the male except for her slightly smaller size and her less robust bill, frequently walked up to her mate and gently preened the feathers on his neck and breast while he stood motionless, bill pointed to the sky, eyes closed, utterly content. Sometimes the two birds crossed their bills, and the female crouched down and with quivering wings seemingly begged for food. After that both gurgled and talked to each other with short *kra-kras*.

On days when the warm air from the lowlands rushed up the canyons, or the winds howled and the thunder grumbled around the upper slopes of Mt. Mikeno, the ravens played their airy games. Like black messengers from Thor, they emerged out of the wind-tattered clouds, rushing downward with folded wings, until abruptly they veered skyward and disappeared, only a disembodied *krrrua-krrrua* echoing from the clouds.

On October 3, I tracked group VI high on the slopes of Mt. Mikeno. It was toward noon, and the gorillas were resting on their backs with the hot rays of the sun burning down on their bare chests. The Virunga range spread glistening to the east. I sat on a grassy knoll with my back against a bluff, my feet dangling over the depth of a dusky canyon. Ahead of me were the gorillas, and beyond them the slope swept upward, the breeze moving the green and silver leaves of the senecios until I was dizzy with the light. While eating my lunch, I spotted the ravens high above me, small black spots against the white of a cloud. When I whistled, which to the birds signified food, they descended in leisurely arcs. The gorillas ducked when the gliding shadows of the birds passed over them, and the male jumped up and roared; females screamed, some looking at me, others at the ravens. Then, as if in play, the ravens swooped at the gorillas, diving low over them again and again. The male grew angrier than I had ever seen him, and the females milled about in utter confusion. The apes obviously failed to find this a game, and the ravens, well-satisfied with their mischief, landed in a heather tree near me to consume the rest of my lunch. They then headed into the valley, but an hour later they returned and again flew in unison at the gorillas. As before, the male stood on all fours and roared at the feathered missiles that hurtled toward and over him.

The ravens sometimes carried bits of dry grass in their bills in October and November, and we suspected that somewhere in a rocky niche on the bluffs of Mt. Mikeno the pair was constructing its nest. During the latter part of January we were in Uganda, and the day after we returned to Kabara, on February 2, the ravens visited us. With them they brought two fully fledged and squawking youngsters. The female still fed her large offspring, but the male was quite intolerant of their presence, chasing them whenever they approached. Once he and a young raven fell out of a tree, each grasping the other's chest with its claws and hammering away at each other's head with the bill. In late February, the young birds left to lead an independent life, apparently driven from their home, for they never returned to the meadow. But the male and female continued to preen and gurgle by our door. I saw them for the last time in September, 1960, still soaring near their mountain home and probably ready to raise another brood.

By virtue of its location between the mountains, our meadow was also a place where various birds came to rest briefly before continuing on into the lowlands. Yellow-billed ducks, singly and in pairs, visited our little pond. They were drab birds, mottled all over with sooty gray, but their bills were bright yellow. On November 24 Kay called to me as I came home: "There's a strange bird by the lake!" Sitting on its haunches by the water's edge was a large, wholly black stork — an open-bill stork, so called for the mandibles of the bill which fail to meet for part of their length. This curious bill is apparently used for extracting snails and mollusks from their shells. The bird remained with us for over an hour, the green and olive of its plumage iridescent in the sun. On another occasion a brief drama unfolded over the lake. A small sandpiper with a white tail flew over the surface of the water. Suddenly an *Accipiter* hawk swooped from a nearby tree and with extended talons tried to grab it. Both birds wheeled and dodged, one after the other, and just as it seemed that surely the end of the sandpiper had come, it dove under the surface of the water, and the hawk flew off. The rarest visitor to our meadow was an Ayres' hawk-eagle, which landed in a tree and regally perched there for ten minutes, the crest on its crown erect, a lean and bold look in its eyes, before disappearing as suddenly as it had come.

One of our tasks was to make a plant collection. After the pressed specimens were identified, I would have a fairly good inventory of the plants which are part of the gorilla's realm. Kay and I meandered over the meadow and along its edge, picking violets, pungent mints, fleshy-leafed stonecrop, and the many other plants that grew hidden among the boulders and in the undergrowth. Most of the more showy blossoms were yellow — the various senecios, the *Hypericum* — and these stood out in striking contrast to the dark-green foliage. Tiny white daisies covered the meadow in front of our cabin like flakes of snow, but our rooster developed a craving for these flowers and within a few days they were gone. Each massive *Hagenia* branch had its own garden of mosses, lichen, and ferns, and orchids with purple blossoms shone like torches in this somber greenness.

When we discovered a clear footprint of a leopard or gorilla in the soft earth, I poured some plaster of Paris mixed with water into the track. A few hours or a day later we retrieved the plaster which had hardened into an exact replica of the track. Only by holding the print of a gorilla in my hand was I fully able to appreciate its size. The knuckle prints — that is, the four fingers of the hand — of a silverbacked male have a width of six inches as compared to the little-over-three-inches of my own hand. The footprint of a silverbacked male is as much as twelve inches long and considerably broader than the human foot. On one occasion I placed a shiny tin can on the trail to mark the site of a plaster cast. A blackbacked male came down the path after I left and, judging from his tracks, it was evident that he passed the can without breaking his stride or stopping to investigate. That the gorilla should show so little curiosity toward this strange and shiny object came as surprise to me. Later, I came to the conclusion that the gorilla in the wild simply lacks the inclination to investigate new things, and that he rarely handles anything solely out of curiosity. In this respect the gorilla differs markedly from man.

Gorillas passed by our meadow regularly, but they never attempted to inspect our cabin. The sole exception to this was Little Adolf, a lone blackbacked male. I first met him on a sunny morning in a clearing near the lake as he was taking a sunbath. He was draped on his back on a log, completely relaxed. When he saw me, he let out a roar and bolted. Yet irregularly thereafter — in mid-November, in April, and in May — he spent several days near our cabin, usually just above us on the slope. Whenever he spotted activity on the meadow, he ranted and raved at us, remaining carefully hidden in the undergrowth. Only after staring in the direction of the sound were we sometimes able to discern the hairy top of his head and a pair of curious brown eyes gazing steadily in our direction. He seemed genuinely interested in watching us. But it was always rather startling to step sleepy-eyed from the cabin at dawn and be greeted by an explosive roar. Although it came from a distance of about a hundred feet, it sounded as if he were at the corner of the hut. Even so, we became very fond of Little Adolf.

Other lone males near the cabin were not equally vociferous. I remember one morning when Kay and I came across the fresh trail of a lone silverbacked male in the forest near the edge of the meadow. The dung was warm, which meant he was just ahead. With extreme caution we traced his route through the undergrowth, one step at a time, every so often climbing onto a stump to scan the vegetation. But the forest ahead was calm and unmoving. Once the powerful odor of the male assailed us, a rather peculiar smell, somewhat sweet and resembling burning rubber. Was he waiting somewhere in ambush? We were both uneasy. Kay has one failing: if anything fearful works on her emotions, her stomach protests. Now, when my senses were keyed, trying to perceive some clue to the location of the gorilla, Kay's insides growled loudly behind me. This invariably made me jump, and we decided to abandon our pursuit.

Andrea and the park guards were not fond of Little Adolf or of gorillas in general. To them these apes were all that is strange

A gorilla nest in the crotch of a Musanga *tree in the Maniema Forest.*

The Lowa River is here too broad to be bridged by a fallen tree, and this prevents the gorillas from expanding their range to the opposite bank.

Kay sits in the fork of a tree and watches group VIII. *Mr. Crest, the silverbacked male, and three other animals are visible.*

Junior, the blackbacked male of group IV, sits calmly only some thirty feet from me.

and fearful in the forest. Once I took Andrea with me to show him gorillas and to prove to him that the animals were amiable. He was not happy to be along, and as the gorilla trail grew fresh he lagged behind. When he had his first view of a gorilla, he showed no interest in it — or for that matter in what I was writing down and why. That he obviously had no interest in the gorillas puzzled me, for surely he recognized the animal as his kin. Many tribes in West Africa express this kinship with the apes, and also with the wise elephant, by believing that apes, elephants, and man have a common ancestor. The Bulu tribe in Cameroun, for example, has this story of creation, reported in the *American Anthropologist* in 1911:

God had five children: the gorilla, chimpanzee, elephant, pygmy, and man. Each was given fire, seeds, and tools, and told to go out alone into the world to settle down somewhere. The gorilla went first, and as he traveled along a forest path he saw some delicious red fruit. After he had eaten his fill and returned to the path, he found that his fire had gone out. So he stayed in the forest, living on fruit. The chimpanzee left home next, and he too became hungry as he passed a tall tree in fruit. He climbed up, and when he returned the fire had gone out. So the chimpanzee, like the gorilla, went to live in the forest. The same fate befell the elephant.

The pygmy went much farther than the others and finally cleared away some underbrush, planted his seed, and built a small hut; he did not cut down the tall trees, but he kept the fire going and learned the ways of the forest. Then at last man ventured forth and traveled very far. He cleared a large garden and cut down the trees, and he built himself a large house. He burned the brush and planted his seeds and lived there till the harvest was ripe.

After a time God went out to see how his children had fared. He found the gorilla, chimpanzee, and elephant in the forest, living on fruit. 'So,' said God, 'you can never again stand before man, but must ever flee from him.' Then God found the pygmy under the trees and he said: 'So you will always live in the forest, but no place will be your fixed abode.' And finally he came to man and saw his house and his garden and he said: 'So your possessions will remain always.'

I am sure that the Africans thought me slightly crazy. Why would I leave the cabin day after day, in fog and in rain, to sit and watch gorillas, without even taking photographs? But to be thought of as peculiar has its advantages, for people tend to accept one's unpredictable behavior and sometimes show sympathy. Often a scientist is distrusted by the villagers because they wonder why he acts as he does and even try to anticipate what is expected of them. They know and categorize agricultural officers who make them grow certain crops; they understand the local administrators who establish laws they do not like; and they tolerate missionaries who prohibit them from keeping more than one wife. The motives of such persons are obvious, but the goals of a scientist are incomprehensible to them. I was apparently judged to be harmless, indulging in a personal idiosyncrasy, for I never had any trouble.

Why indeed study gorilla? Why spend thousands of dollars and several years in watching an ape in the wild? One answer was given by Sprat in 1722:

But they are to know, that in so large, and so various an Art as this of Experiments, there are many degrees of usefulness; some may serve for real, and plain benefit, without much delight: some for teaching without apparent profit: some for light now, and for use hereafter; some only for ornament, and curiosity. If they will persist in condemning all Experiments, except those which bring with them immediate gain, and a present harvest: they may as well cavil at the Providence of God, that he has not made all the seasons of the year, to be times of mowing, reaping, and vintage.

For years monkeys and apes have been studied in captivity. Their minds have been probed to determine how well they can discriminate between objects and how rapidly they can learn to do various tasks. Psychologists have watched the interactions of primates in cages, and they have traced the development of infants from birth to adulthood. Yet none of these investigators knew the natural life of the animal they were studying; they had no intimation of how typical the captives were in their development and behavior. To make intelligent and meaningful interpretations from captive animals it is first necessary to know about their natural mode of life.

Monkeys and apes, because of their close relationship to man, are also highly useful in the study of human diseases. For example, each year some 200,000 to 300,000 monkeys are imported into the United States at a cost of twenty million dollars for use in the production of polio vaccines and research in various medical fields. As always, man is harvesting ruthlessly without thinking about future supply. Once it was generally assumed that the number of rhesus monkeys in India was unlimited. Yet in spite of the fact that thousands of rhesus monkeys were trapped yearly for shipment to the United States, their status and behavior was completely unknown until Dr. C. Southwick studied them in 1960. He came to the conclusion that the supply of wild rhesus monkeys was diminishing rapidly and that critical shortages were apt to develop. The chimpanzee is more closely related to man in its blood types and susceptibility to certain diseases than any other animal. More and more of these apes are being shot and trapped in the wild for use in medical research. No one knows how many there are left and how many may be taken from an area without exterminating the population. No primate can long survive sustained and uncontrolled persecution. Unless rhesus monkeys and others are bred in sufficient quantity in captivity or taken on a sustained yield basis in the wild, medical research will suddenly find itself deprived of animals in future years. To manage a species intelligently in the wild or in captivity, a knowledge of its life history and social behavior is essential.

There is another reason for studying animals, to me the most important of all: that man learn to understand himself. In an age when man is increasing his mastery over nature and is occupied with devising new weapons with which to obliterate himself, he is as yet only on the threshold of learning the causes underlying many of his actions. Because the behavior of man is so masked and modified by the culture into which he is born, it is often easier to obtain a clearer perspective of some problem of human behavior by studying other animals. The closest relatives of man,

and a group to which he belongs, are the primates — the lemurs
and other prosimians, the monkeys, and the apes. Clues to the
origin of human behavior and human society might thus be found
in various animals and especially in the primates. As Carl Sandburg once wrote:

There is a wolf in me . . . fangs pointed for tearing gashes . . .
    a red tongue for raw meat . . . and the hot lapping of blood —
    I keep this wolf because the wilderness gave it to me and the
    wilderness will not let it go.
There is a hog in me . . . a snout and a belly . . . a machinery for
    eating and grunting . . . a machinery for sleeping satisfied in the
    sun — I got this too from the wilderness and the wilderness will
    not let it go.
There is a baboon in me . . . clambering-clawed . . . dog-faced
    . . . yawping a galoot's hunger . . . hairy under the armpits . . .
    ready to snarl and kill . . . ready to sing and give milk . . .
    waiting — I keep the baboon because the wilderness says so.
O, I got a zoo, I got a menagerie, inside my ribs, under my bony
    head, under my red-valve heart — and I got something else: it is
    a man-child heart, a woman-child heart: it is a father and mother
    and lover: it came from God-Knows-Where: it is going God-
    Knows-Where — For I am the keeper of the zoo: I say yes and
    no: I sing and kill and work: I am a pal of the world: I came
    from the wilderness.

A trivial example may help to illustrate how by watching
a gorilla we can obtain clues about our own actions. Dr. N.
Tinbergen and other students of behavior have found that in
some situations animals may have two incompatible tendencies
or impulses: for example, to attack an opponent and to flee from
him. These conflicting tendencies generate tension, which may
find release in behavior that at first sight is completely irrelevant
to the situation in which it occurs. Such behavior has been termed
a displacement activity. Sometimes, when I encountered gorillas,
the apes were obviously vacillating between the tendency to flee
from me and the tendency to approach more closely, out of
curiosity or perhaps to chase me away. If these opposing impulses
were so well balanced that they canceled each other, the animals

neither advanced nor retreated but stayed in one place. But they quite commonly exhibited two kinds of behavior which were certainly not relevant to the situation: some began to feed very intensively and rapidly, and others scratched themselves vigorously. Both these acts were undoubtedly displacement activities which served to reduce tension in the animal. Man, in situations where he is undecided about a course of action or unsure of himself commonly eats something or takes a drink, and very often scratches the hairy remnant on the top of his head, actions which are readily understandable if one has watched the apes.

At least once a week I walked to the Bishitsi bluff, two miles to the north, to see if any gorillas had passed through that part of the forest. I used the heavily worn buffalo trails that wound through the undergrowth, taking an unexpected turn here, bypassing a certain ravine there. Some of these forest highways followed the contours of the slopes, worn deeply into the soil by the pounding of innumerable hooves. But many seemed almost purposeless in their meanderings, and yet they were used again and again by the animals — perhaps traditional routes handed down from generation to generation. One day the guard and I cut a narrow trail through the stands of lobelias and nettle fields directly to Bishitsi. Soon the buffalo found this new route, leopards padded along it, and park guards used it on patrol. A new forest path had been created, and I wondered if it would persist, if years hence the animals would still travel over it.

The terrain toward Bishitsi is gently rolling, pleasant for walking. There are no biting insects — mosquitoes, black flies, ants — and no snakes or other reptiles except two species of chameleon, nothing to hurry the step, nothing to prevent one from peacefully sitting down anywhere. Only the stinging nettles prevent it from being too perfect. The nettles were the most virulent I have ever known. In places where the soil was well drained, the canopy broken, and the surface fairly flat, the nettles grew in dense stands, six to eight feet high, and covering acres. They were innocuous looking plants, with spindly stems and leaves covered with tiny white hairs, but when they touched the skin they burned with unabating fury. After a sojourn through a nettle field, my legs were red and swollen and red welts covered my face and hands. Yet a plant which could raise a welt on my skin through two layers of cloth did not halt the gorillas. The apes walked through nettle fields without hesitation, used the plants in nest construction, and ate the leaves and stem, handling and chewing the hairy portions with apparent impunity. I wished that I knew their secret.

There is a gnarled tree at Bishitsi with a huge limb that grows out over the edge of the bluff. I liked to sit there on a cushion of moss, letting my eyes roam over the forest far below. Clouds frequently blew in from the north and collided with the bluff. Tatters of white mist sprayed over me like the breaking of the surf. Once two squirrels chased each other in a tree adjoining mine, scrambling around the trunk, their golden sides and grey back a blur. They raced over the ground and up on the branch on which I was sitting, quite unaware of me. One jumped on my shoe and looked back at its playmate, which braked to a stop and stared at me with its shining black eyes before scampering off, followed by the other.

While sitting on the branch, I could hear the courtship calls of the olive pigeon, a heavy-bodied bird with bluish-grey plumage, a white-speckled chest, and a yellow bill. These pigeons were shy creatures, sitting quietly in the dense crown of a tree until suddenly they took flight with such a loud flapping of wings that I was startled. The call of a displaying bird was a resonant *bruuuu*, rising in pitch and followed by several rolling, deep *bruuu-ooo-ooo*. Then the bird flew high into the air and sailed downward with arched wings and spread tail, emitting a strange donkey-like braying. I found six nests of this pigeon between November 19 and April 27, all flimsy structures consisting of but a few dry twigs placed into the crotch of a tree within twenty feet of the ground. Each nest contained only one white egg or one squab covered with pale yellow down.

Another route which Kay and I followed frequently led to the Rukumi meadow. We were happy now and then to leave the forest for the wide horizons of the grasslands. The luscious blackberries that ripened on the meadow throughout the year were often our specific goal. Sometimes we took along containers and collected several quarts of berries which Kay later used in pies and turnovers and made into syrup for pancakes. At other times our steps took us simply from bramble patch to bramble patch as we plucked and ate the ripe fruit which we both craved after our steady diet of tinned goods. Gorillas too have a predilection for these berries, and once I came across a silverbacked male at the edge of the meadow, leisurely plucking the berries and popping them into his mouth.

On many days, while I was searching for and watching gorillas, Kay remained at Kabara, filling her time with sewing, baking, and all the other tasks of a housewife, and she also handled all the correspondence of the expedition. Here is how she describes an average day:

"Standing on a rise by the cabin, I would watch George cross the meadow and wave to me as he reached the edge of the forest and disappeared among the trees. Then I would retreat to my chair by the stove, pour a cup of coffee, and contemplate my various tasks. My days were evenly divided between keeping the stove and Andrea going — both preferring to smoke leisurely. The coffee finished, I would call Andrea and try to jar him into activity. Each day he was expected to cut wood. But he did not like going alone into the cold, wet forest, and so welcomed any delay, such as digging us another pit for the garbage or sometimes killing and cleaning a chicken, which I then simmered all day to make it tender. If the sky looked as if it would remain clear for several hours and if the rain barrels were full, I had Andrea wash clothes. Since the water in the barrels was bitterly cold, I would add one kettle of boiling water to the tub, more for psychological reasons than for anything else. No matter how many clothes were being washed, the job took about an hour and a half; it took as long to do one pair of trousers, a sweatshirt, and three pair of socks as it would for three or four times that amount! Each item was lavishly soaped and scrubbed — soap being used in such quantities that I finally had to ration it. At first Andrea seemed unable to understand how our clothespins (the spring variety) worked, so I usually hung the clothes for him and this also served to speed him up a bit. I think he must have practiced, however, for once, when weary of waiting for him, I took a short walk, and when I returned he had hung them himself, though awkwardly and with several pins per item.

"Sometimes I sat nearby as Andrea squatted by the wash-tub, and he and I attempted to learn a bit about one another's lives. I always had to go into the cabin for the Swahili dictionary, but after considerable difficulty we would manage a brief conversation.

" 'Does Madam have a boy in America?'

" 'No. In America I have machines to wash clothes and to clean.'

" 'Uhhhhhh.'

Or, another time:

" 'Do you have a wife Andrea?'

" 'No, Madam. Cost many francs.'

" 'How many?'

" 'Three thousand francs, Madam. How many francs in America?'

" 'No francs in America. Wife in America is free.'

"This was followed by an amazed 'Uhhhhhh!' and finally,

" 'Will Bwana and Madam take Andrea to America?'

"After prodding Andrea, I began my own tasks. First, shake the sleeping bags and perhaps air them outdoors if the day was nice. Then, having added a bit extra in the way of feathers and dust to the accumulated dirt on the floor, I would take the broom and sweep everything very conveniently down the many cracks between the floorboards. At one time we collected lichen to stuff between these cracks, hoping to cut down some of the cold drafts, but Andrea always carefully picked them all out whenever he did the sweeping. Then there were the lanterns, which I preferred to clean, refuel, and repair myself, for they were far too precious to leave to our somewhat awkward Andrea.

"Sewing took up more time than I liked. Each of George's socks was redarned many times, and his pants, parka, and even packsack were constantly in need of repairs. I think I must have 'remade' his favorite, dark-green parka at least twice. Even with my thimble I could not manage to get a needle through the webbed straps of the packsack, so I had to push it through by resting it against the table and pushing the material down with both hands, usually puncturing my fingers several times in the process.

"Baking was a special treat, not only because of the end result but also because it was fun to experiment with new ingredients and 'make do' with the supplies on hand. The high altitude made many changes necessary, and the eccentricities of our poor camp oven only added to the adventure. The oven had to be balanced over the primus, which in turn was balanced on the uneven floorboards, with various bits of paper or wood propped beneath the three legs to make it level. Once I tried putting the primus on the large table, but the oven fell off when the table was jarred by our walking. After many trials I found the most level spot on the floor, but we still had to walk softly, and preferably not at all, to prevent accidents. The oven leaked heat at every possible spot, with the result that everything was really only stewed from the bottom and dried from the top. But no matter how doughy or dry, to us everything tasted superb — George never appreciated me more than on baking days. We had berry or apple pies, apricot tarts, turnovers, gingerbread, corn muffins, coffee cakes, peach cobbler, and bread. Even that awful oven could not ruin the bread, usually molasses-oatmeal or cracked wheat, and of all the things I baked it was our favorite. I made one loaf a week and tried to time its baking so that it was ready when George returned from the gorillas. We loved it hot from the oven and spread with butter and honey. The first day we allowed ourselves three slices each and then two slices a piece for two additional days.

"Periodically I would be interrupted by a guard who wanted 'dawa,' medicine. We usually gave the guards aspirin, for which some developed a great fondness. They returned time and time again, occasionally several times a day — they were apparently firm believers in preventive medicine. To a persistent 'case' I finally gave a dose of Epsom salts, telling the 'patient' to drink it very slowly. This always cured them. The guards were also very fond of bandages and would even ask me to put one on an ancient scar. But I would laugh and shake my head, and they would grin in return as if they had been joking—perhaps they were.

"Once the housework was done I might sit at the typewriter to pound out a few letters home or reports for George. Andrea was fascinated by the typewriter, and he often stood quietly behind me to watch. Or I might go outdoors to collect plants or take photographs, or just sit in the sunshine, if there was any, and read and knit, trying to ration the books and yarn so that they would last for the duration of our stay.

"The two white-necked ravens came often and made excellent companions. I had never before tamed anything wild, and this perhaps added to their appeal. After feeding them for some weeks, they became very demanding—cawing loudly as they arrived and continuing to call until I stepped out and gave them some food scraps. Occasionally I went out empty-handed, and then they would peck at my shoes or clothes as if trying to persuade me to get them something. They were terrible thieves, and George and I learned not to leave any of our gear unguarded. Several times I had to rescue Andrea's and the guard's clothing or my soap from the birds. But they were fun and lovable and provided me with many happy and interesting hours.

"About mid-afternoon it was time to think about dinner. I would go into the storeroom and look at all the cases of tinned goods. (George loved to putter in the storeroom arranging and rearranging the supplies—they made him feel like such a good provider.) For some reason it was easy to forget what we had eaten the past few evenings, and I would try to choose according to the number of cans left in the open cases. Eventually I made out menus for an entire month in order not to end up with a steady diet of corned beef. The meal was planned around the meat we were eating, and although we had a reasonable variety— stewed steak, steak and kidney, ham, hamburger, ox tongue, salmon, corned beef—most of it tasted the same, and I used lots of onions, mushrooms, various herbs, and wine to disguise the flavor. We generally had a soup, meat, vegetables, potatoes or rice, and for dessert a little fruit or something baked. We preferred the "one-pot" meals—Spanish rice, goulash, spaghetti, stews. Our favorite was beef stew, made from potatoes, onions, leeks, tomatoes, peas, mushrooms, some dried vegetables, carrots, a bit of Worcestershire sauce, various seasonings, and red wine. I had found a good recipe for herb dumplings, and George invariably asked me to make them.

"By about four o'clock I could begin to expect George home. Every so often I stepped outside the cabin to look toward the trail. Finally he came, chilled and hungry, and I would give him hot cocoa or bouillon, and if I had baked he was given an 'extra.' Then while he got into a warm sweater and dry pants, I hung his sodden clothes by the fire and sat beside him to hear about how the gorillas had been and what he had learned that day."

Our days in the mountains sped by, days without time, month after month, with little to break the steady rhythm of life. Here near the equator the length of the day changed only imperceptibly, and the temperature remained the same throughout the year; there were no regular seasons in these mountains, except that some months were wetter than others. In all our time at Kabara we had only one visitor, for the park authorities were strict in forbidding access to the volcanoes. Only Grenold Collins, an old friend of the family, stopped by for two days in late October. It seemed strange to sit and talk to a third person about the outside world which to us seemed so far away. Our lives were circumscribed by the hut, meadow, and forest, and the creatures in it; our conversations had ceased to stray much beyond this realm. The gorillas were our neighbors, and we gossiped about their daily doings: Mrs. Patch has had a quarrel with Mrs. Blacktop; Mrs. Callosity Jane let her baby ride on her back for the first time; the injured eye of Mrs. Bad-eye looks worse. Living in this close and continuous intimacy with each other, Kay and I had learned to anticipate each other's actions and the details of our routine. Words became more and more unnecessary as an aid in regulating our lives. I found it easy now to understand how gorillas co-ordinated their actions so successfully by gestures and limited vocalizations, and I realized how completely adequate their mode of communication was for this simple way of life. Only once in a while Kay sighed: "I wish I had a woman to talk to — about recipes and clothes, all the things that don't interest you."

We had no radio at Kabara, for we were loath to disturb the tranquillity of our life by introducing noises from the outside world. From the park authorities we never received news, except that once every three weeks or so, with the changing of the guard, our mail was delivered. Mail day. On the expected day, Kay peered out the door every few minutes with the hope of seeing the new guard emerge at the edge of the forest. Sometimes the guard was sent up several days late, but sooner or later he came, accompanied by a villager who carried his three-week ration of beans, rice, dried fish, and beer. It was always with excitement that we tore open the letters from home and read *Time* or *Newsweek* to find out what was happening in the Congo. Then, after a few hours, the outside world receded from our minds again until another mail day.

To the guards the stay at Kabara was a punishment, a life which they loathed. It was cold and damp, and except for Andrea there was no one with whom they could talk and drink beer. Kay was always somewhat fearful of having me roam the forests alone all day and sometimes all night. When she herself did not accompany me, she urged me to take the guard. To please her, I made the guard accompany me about half the time, but when the gorillas were near I went on alone, leaving him behind to wait several hours until I returned. Most guards understandably disliked these journeys with me intensely, for the nettles burned their legs and they feared the animals of the forest with an irrational fear that seemingly no amount of exposure could allay. Once the guard Donati, a tall, cocky fellow, hiked down a trail with me. A buffalo, apparently disturbed from his slumber, crashed through the underbrush in our direction. I stepped behind a tree, but Donati raced screaming down the trail as if the buffalo were in full pursuit. And the buffalo lumbered by without changing direction. Similarly, Desmos, a thin wiry guard with a touch of Watutsi, was directly behind me when inadvertently we stumbled on a group of gorillas. The male reared up from the vegetation and emitted a series of shattering roars.

Desmos flew down the slope, and I did not see him until several hours later when he limped back with a wrenched knee, the result of his precipitous retreat. The name "guard" seemed, on the whole, to be honorary, with little relation to the function they usually performed while with me.

Only at intervals of weeks and sometimes months were the days of watching gorillas broken by other events. There was, for example, Christmas Eve. The day before we sent Andrea and the guard home to their village, and now we were alone. The weather was dismal, with dense fog that came rolling, rolling over the meadow, a translucent gray mist in which the forest faded and reappeared as if in a dream. From the lake a well-worn buffalo trail led up the slope of Mt. Mikeno, and I took this route in search of a Christmas tree. Far up on the mountain, where the *Hagenia* ceased to grow, I found the trees I was seeking. Stands of heather covered the ridge, and after a little searching I found a tree about five feet high with fairly symmetrical branches. I cut it down with my knife and started homeward, the tree on my shoulder. Soon I was running down the mountain, through the dark groves of trees, through the fog, with the wind rushing by me and the saplings lashing my face, and I only hoped that I would not collide with a buffalo. Down I hurtled, almost flying, my whole being filled with exhilaration, until I reached our lake at the base of the ridge. Kay was waiting for me, and together we decorated the tree with yellow blossoms picked in the forest and with silver stars cut from foil. To surprise Kay I had brought our favorite tree ornament from home, a little wooden carving of a chickadee, and this we now unwrapped and placed on a bough. Beneath the tree we spread the presents of books and clothing, and on the table nearby was a bowl filled with oranges, nuts, and a pineapple.

A drizzling rain fell steadily on Christmas Day. Toward noon I had a sudden urge to visit the gorillas, and I set off toward Bishitsi. Even though I held a small plastic tarp over my shoulder, I was soon soaked through. After an hour of tracking I came upon group VII moving very slowly in the rain, paying little attention to me as I climbed into the low crotch of a tree and lay down on a massive horizontal branch. Fog swirled in among the trees, and the drizzle changed to a downpour. Slowly the gorillas came toward me and sought shelter against the trunk of a large tree adjacent to the one in which I was lying. The long hairs on the bodies of the apes were sodden, and a faint odor like that of a wet blanket emanated from them. They huddled motionless under the tree, bodies pressed together apparently for warmth. Hours passed. The rain continued to pelt down on my back, and the gorillas remained in their shelter watching me mutely. There is a tale that on Christmas night man and animals forget their differences and can converse with each other on equal terms. The gorillas talked to me at times with their expressive eyes, and I felt that we understood each other, but no sounds passed our lips. The forest darkened imperceptibly, and with a shock I realized that it was five o'clock. I left the gorillas and hurried home. When I emerged from the forest at the far end of the meadow, I saw Kay standing on the rise by the cabin, waiting for me, looking very lonely in the fog and gathering dusk. Tears were streaming down her face, and when I held her close, she told me that it was very late, it was Christmas, and she thought that I had been injured because I had not come home. We had a big dinner that night — chicken (the last of our flock), asparagus, corn muffins, rice, and an almond cake. We had some wine too, and at this altitude a single glass was enough to make us aware of it.

Earlier in the month, on December 11, I had gone to Bishitsi, and scanned the plains below as I usually did. Herds of cattle wandered into my view, and kraals and the huts of Watutsi herders stood where only a week before the undergrowth had stretched unbroken. Why this sudden invasion of the park? Why did the authorities do nothing about it when surely they were aware of the situation? Cattle and wild life cannot coexist for long in these mountains, for the domestic stock tramples down and grazes the plants so thoroughly that other species like buffalo and gorilla are unable to subsist on the remains. I knew something of the constant clashes between the arrogant Watutsi and the park authorities. The Watutsi needed new grazing lands for their large herds of cattle. But the Belgian administrators refused to help the park by levying substantial fines against the Watutsi when they were discovered in the park.

Yet this time there was another reason for the sudden appearance of the Watutsi. They were refugees hiding out in the forest with their cattle. Civil war had broken out at the foot of our mountain, with the Hutu tribe fighting against their former overlords, the Watutsi. Thousands of Watutsi fled into Uganda and crowded into missions for protection, while the Hutu rampaged in bands, pruning the tall Watutsi down to size, so to speak, by cutting off their feet. Our first news of this did not come from the park authorities, but from *Time*, which was delivered to us a few weeks later. The guards chased the cattle away from Bishitsi, back over to the Rwanda side, but I wondered how many other herds were wandering in the area away from the regular routes of the park patrols. I had not yet been on Mt. Visoke, 12,200 feet high, which straddles the Congo-Rwanda border, and I decided to survey this section of the park for gorillas and for cattle.

The guard and I left Kabara early on the morning of January 5, each of us carrying enough food for three days. The first part of the journey northward to Bishitsi went rapidly, but then we headed east through an area heavily overgrown with nettles. There were few buffalo trails, and we cursed the undergrowth as we hacked our way along. After some two hours we emerged on a wide cattle trail that led toward the Rwanda border. At the base of Mt. Visoke we came on two meadows, and there beneath some trees we set up our camp for the first night. While the guard stretched a tarp between some saplings to keep us dry in case of rain, I reconnoitered along the edge of the meadows. Suddenly, just ahead, I heard the clanging of cow bells, and two Watutsi, dressed in long overcoats and carrying spears, drove twenty-five head of cattle out on the grassland. When I told my Hutu guard about the Watutsi, he was understandably scared, and we sat silently in the undergrowth until evening, when the fog rolled in and we dared build a fire and cook our meal.

With the first light of day we broke camp and angled directly up the slope of Mt. Visoke. Below us, at the edge of the meadow, burned two fires with several Watutsi crouched around them. From the safety of the slope, I shouted over the forest in Swahili and French as if giving orders to my troops: "All right, men. Surround the Watutsi! Quick, quick! Catch them."

Thinking themselves attacked, the Watutsi bolted into the forest, leaving their meager belongings by the fire. I was pleased at the result, but the guard frantically motioned me to hurry after him up the mountain. The climb up Mt. Visoke was easy. Soon the forest gave way to scattered giant senecios, and then we reached the flat summit, which was first ascended by the geologist Kirschstein in 1908. Gorillas had preceded us to the top, but we did not see them there, nor much of anything else, for dense fog engulfed us and a biting wind drove into our faces. Occasionally the clouds parted briefly to reveal a crater lake, some quarter mile in diameter, far below us. We followed the rim of the crater, walking a tightrope through the clouds, and then began our descent of the mountain on the opposite side. Hundreds of cattle were visible as small brown dots on the more open terrain toward Mt. Sabinio. The saddle between Mt. Visoke and Mt. Sabinio is covered with grass meadows and stands of bamboo, ideal, unfortunately for cattle grazing. I had visited this area in July, 1959, at the height of the dry season, and found few signs of gorillas, who prefer forests with a more succulent vegetation. Too many Watutsi roamed through the forest around Mt. Visoke, making our travel unsafe, and we decided to return to Kabara, arriving thoroughly exhausted that evening.

After we had been five months at Kabara, our food supply was nearly gone. On January 16, the porters came to fetch us, and two guards remained behind to keep their eyes on our belongings until we returned. For two weeks we talked and traveled. But the need for solitude had grown on us imperceptibly, and after a few days of civilization we thought longingly of our mountain home. We had become more introspective, the need for constant outside stimulation had left us. Even the many-coursed dinners at the hotels to which we had once looked forward with anticipation soon palled on us. On February 1, in pouring rain, we returned to Kabara. It was good to be back.

Later we made two more brief trips down the mountain, one in late February to visit with Dr. Fairfield Osborn, the president of the New York Zoological Society which sponsored our expedition, and his wife, and the other in mid-May to spend several days with Kay's mother who was making a world tour. When Kay and I returned alone to Kabara after the second trip we strayed into a herd of elephants that stood scattered and motionless among the crowded ranks of bamboo stems. We smelled them, heard their stomachs rumble, and one animal trumpeted. Yet we did not glimpse them. We tried to circumvent the beasts by clambering down into the Kanyamagufa Canyon and following the course of the rocky stream upward. There, under a rock overhang, we came on a fire and the remains of a duiker — apparently poachers had heard our approach and fled. We reached the trail again after a hard climb up the canyon wall and almost collided with an elephant that stood around a bend in the path. After another laborious detour, we were only too glad to leave these formidable beasts behind.

When we began our study Kay was understandably concerned about the elephants. "They never go up to Kabara," I assured her, for I had seen no sign of them during our first visit in March, 1959. But when we arrived in August, there in front of the hut was a huge pile of elephant dung, deposited no doubt by a straggler since no other elephant ascended to these heights during our stay.

Each time, after a visit to the lowlands, our life easily fell back into the familiar routine as day after day I gathered more data on the gorillas. Much of it was quantitative, more of the same thing, behavior that I had seen many times before. At times it seemed that it was hardly worth my time and effort to watch the gorillas feeding and resting yet again, but only through constant repetition was I able to discern and interpret many subtle actions which at first I had overlooked.

When for the first time I set eyes on Mt. Mikeno, I knew that

I would have to climb the peak. But it looked like a difficult undertaking, and everyone assured me that it was. The first attempts to reach the summit had all ended in failure. In 1908 the Duke of Mecklenburg reached the base of the huge rock face that led to the top. Derscheid, a member of the Akeley expedition, ascended to within two hundred feet of the summit in November, 1926, before being driven back by foul weather. Early in 1927, a similar fate befell the ornithologist J. Chapin and two missionaries. Finally, in August of that year Père van Hoef and Père Depluit, accompanied by a Mr. and Mrs. Leonard of Belgium, reached the barren top, a feat which was repeated in 1938 and at least three times since then. A favorite spot of mine was at an altitude of thirteen thousand feet, at the crest of a ridge that plunged down to the Kabara meadow. I liked to lie on the soft carpet of *Alchemilla*, letting my eyes roam over the distant hills and the forest and the sky, letting the peace of the mountains flow into me. When I left the cabin to escape to these heights Kay always chided me, somewhat annoyed: "You are alone with the gorillas all day, and then you come home and say you need to be alone again!" For some unexplained reason a day with the gorillas often had the same effect on me as a crowd of people: I needed to be by myself for awhile. After hours of scanning the rock wall of Mt. Mikeno from my seat in the sky, I learned all the ridges and cracks and found the best route to the summit. It was March, the weather fairly good at least part of the day, and I knew that if I was going to climb the peak I would have to do it soon.

March 13. The sun had not yet risen, and the temperature hovered around the freezing point when the guard and I left Kabara to begin the ascent of the mountain. An hour later we emerged from the forest into the senecio zone, sweating now from the effort of the climb. We followed a buffalo trail upward to the base of the huge rock wall that rose sheerly to the peaked summit about fifteen hundred feet above us. The sky was clear, but far in the distance, past Goma and out over Lake Kivu, white clouds balled up. I left the guard sitting on a knoll to await my return and continued alone. I traced the face of the rock wall until at one point its slope became more gentle and covered with moss and stunted senecios. I sank my hands into the icy moss to find handholds and slowly pulled myself up, a step at a time. My fingers soon were numb from the cold. The face of the mountain became a jumble of precipitous ravines, burnished by golden moss that cascaded down the walls. It was a treacherous moss that occasionally broke off suddenly under the weight of my body. But slowly I ascended, pulling myself up the citadel of rock until I reached a gentle plateau, covered with sedges and a barren rounded mass of smooth and slippery earth. Hailstones were scattered about, and, now that the sun warmed the mountain, mist rose all around me. It was completely silent, and I found no evidence of mammals or birds as I crawled up the final rise to the rocky summit, three and a half hours after leaving Kabara. There was a broken bottle and the lid of a can at the highest point, but no cairn or notes. I wanted to remain on the mountain to enjoy the view, Mt. Nyiragongo sending up its mushroom of steam and the blue hills of Rwanda stretching to the horizon, but a solid wall of clouds had reached the base of the Virunga Volcanoes. Rapidly I followed my own footprints down the mountain, grateful that they were still visible in the dense fog that now engulfed me. Six hours after the guard and I set out from Kabara we were back home, and thunder rumbled around the summit.

At dusk I liked to stand on a little rise by the cabin. Andrea and the guard were in their shed talking, and the smoke from their fire crept out the door to linger briefly over the meadow like the night's first reluctant ghost. The faint light of our kerosene lantern glowed softly in the window. I heard Kay puttering around, preparing the evening meal. All around me was the forest, silent now that the sunbirds had ceased to sing. Only the olive thrushes continued to scold. Sometimes a francolin (*Francolinus nobilis*) shattered the stillness with its noisy assertive call — *tisokwe-tisokwe*. These reddish grouse-like birds were solitary and shy, and I rarely saw them, for instead of flushing at my approach they scurried away through the undergrowth. After the francolins and thrushes quieted down there was no sound and no movement except for the rapid flight of an occasional small bat and the hanging lichen on the branches swaying softly.

Dusk was the time when the black buffalo became most active. Although herds of up to twenty animals grazed on the Rukumi meadow and in the bamboo, those in the *Hagenia* woodland tended to be very small — groups of two to five and many lone bulls. The grass on our meadow was tender and green, and almost daily a buffalo or two came and grazed around the hut. When disturbed they stood with head lowered, then wheeled about and raced away, muscles rippling under their shining flanks and tail standing stiffly horizontal. With pounding hooves and a rolling, harsh, *bruuu* call, they headed for the cover of the forest. These buffalo were gentle animals that never sought to attack. But their size and the curve of their horns commanded respect, and at night we were most careful not to bumble into a grazing animal. After dark I usually urinated near the cabin, inadvertently creating a salt lick. It was not long before the buffalo discovered this salt-soaked earth. From then on, almost every night, we heard one of these huge beasts paw the ground, snuffle, lick, and smack its lips, and finally rub its hide against the cabin wall.

The evening meal was the high point of our day. With pleasure I wolfed down whatever Kay had prepared, sitting by the stove, trying to keep myself and the food warm. After dinner Andrea washed the dishes, his last chore of the day. Andrea always moved in a dream world of his own, rather slowly and seemingly weary. While drying the dishes he stopped now and then and stood perfectly still, gazing at the board wall in front of him as if overcome with the beauty of the view. Andrea was slow and lazy, but he was congenial and honest, and we were happy that he was willing to remain with us at Kabara.

After dark Kay and I walked out on the meadow. The sky had cleared, and the stars and moon spread a thin layer of silver over everything. Occasionally an owl hooted far away. Once Kay answered its mournful *hoo-hoo*. Back and forth they called to each other, the owl drawing nearer, until with silent wings it landed in the tree above our heads. The small scops owl gazed into the beam of our flashlight, its yellow eyes unblinking and the feathery horns on its grey and buff head standing very erect.

At certain times of the year the forest became alive at night with the unearthly screams of tree hyraxes — screams that sounded like the croaking of gigantic frogs, the laughing of hyenas, and the wails of a woman being choked to death. Hyraxes, the coney of the Bible, are rabbit-sized creatures with soft grey-brown fur, small velvety ears, and a blunt black nose like a shiny truffle. Each foot has only three toes, the tip of each covered by a nail. Hyraxes are rather solitary animals in the forest, with the habits of rodents, living in the hollow trunks of trees and in burrows beneath the boulders of lava rock. Their nearest relatives are the elephants, a fact hard to believe until one examines the shape of the skull and the curious upper front teeth which consist of two miniature, three-sided tusks. The calls, it has been said, are given by males to attract females. Usually one animal begins to moan and screech, between eight and eleven in the evening, and soon others join in until the mountains vibrate with their calls. An hour later all is silent again, but between one and four o'clock in the morning the hyraxes often resume their chorus. I could find no pattern to these calls, for they occurred at irregular intervals, perhaps for two or three days in succession, and on rainy nights as well as on clear ones. We heard them in August and September and rarely in October. For four months, from November through February, not a hyrax wailed, but in March and April they began again, and in May their sounds were common.

Before going to bed, we huddled close to the stove, sipping a cup of hot cocoa. Water boiled in the kettle, and above our heads hung my wet clothes. We sat without talking, soaking in the heat until the coldness of the night settled on our cabin. Then we prepared for bed, usually by eight o'clock. We washed our face and hands in rather cursory fashion — one simply does not become dirty in the wilderness. Occasionally, at intervals of several weeks, we bathed in our galvanized iron tub. Then, hurriedly we crawled into our sleeping bags, lying rigid and shivering until our bodies had taken the chill from the bedding and we drifted off to sleep.

CHAPTER 8

A Gorilla Day

By midafternoon the gorillas' day has consisted of sleeping, feeding, and sleeping, in that order, and the activity next in sequence is obviously feeding, which they continue until dusk. They forage leisurely, with much sitting around, alternated with occasional bursts of travel at two to three miles an hour. As the forest grows dusky, their movements become very languid, and they collect near the dominant male. If something, like my presence, disturbs them at this time, the animals immediately become restless and move away to bed down for the night far from any prying eyes. But if everything is peaceful, they sit around indecisively, seemingly waiting for someone else to make the first move. Then, about six o'clock, or as early as five o'clock if the sky is heavily overcast, the leader of the group begins to break branches to construct his nest, and the other members follow suit.

The rains came in October. Day after day the dark clouds covered the peaks. Fog rolled into the saddle, not on little cat's feet but ponderously, like on an elephant's. It swirled about, carrying with it the faint odor of moldy leaves, and brought a drizzling rain. In October, November, and December we averaged twenty-three cloudy days a month and about two and a half hours of rain a day. Every leaf dripped, the lichen and moss were like sponges engorged with water, and the trail was a morass, with puddles standing in the buffalo hoof prints. Occasionally, when I stood in the sodden forest, muddy from the waist down and with

all my clothes wet through, it fleetingly occurred to me that somewhere there must surely be an easier and climatically more in-
viting place to study. But then I looked about me and saw the
massive bases of the mountains, and the forest of gnarled trees
with a light like muted silver filtering through the tresses of grey-
green lichen that festooned the branches. This Elysium was mine,
shared only with Kay and the gorillas and the other creatures;
"climb the mountains and get their glad tidings, and nature's
peace will flow into you . . ." wrote John Muir, and I knew what
he meant. I would not have traded Kabara for any place on
earth.

The gorillas did not seem to enjoy the rain. At the beginning
of a downpour, those animals in trees descended to sit hunched
on the ground, and infants returned to their mothers, huddling
against them to keep dry. Sometimes two juveniles clasped each
other to the chest and remained motionless as the water streamed
off their bodies. At such times the gorillas seemed withdrawn from
the world, their animation suspended, and even my appearance
was usually not sufficient to rouse them. Once I reached a group
during a storm, and no gorilla moved even though I cowered in
the shelter of a tree within twenty feet of the nearest animal.

The response of gorillas to rain puzzled me. More often than
not, they remained in the rain when they could easily keep dry by
moving a few feet to sit under the leaning bole of a tree. Occasionally, however, the whole group crowded into the shelter of a
tree with much pushing and shoving as each animal juggled for
the best position away from the drips. Once a female and a juvenile tried to crowd under a tree already filled to capacity, but
the newcomers were ejected with screaming and baring of teeth.
Yet a group which sits in a shelter one day may remain in the
rain on another day. I never could fathom the reason for this in-
consistency: perhaps gorillas simply did not care enough one way
or another. The same thing happened at bedtime. In one instance,
all the members of a group were snug and dry under trees and
fallen logs, but they left their havens to build nests under the
open sky where they received the full force of the rain.

Gorillas usually walk on all fours, and when they do move

bipedally the distance rarely exceeds five feet. On two occasions when a gorilla walked over twenty feet in an upright position, it was raining, and the animal seemingly hesitated to wet its hands and chest on the vegetation. A silverbacked male once walked bipedally with folded arms some twenty-five feet in the rain, and a female hiked sixty feet to the shelter of a tree.

During the periods of heavy rain, the gorillas seemed to catch colds more easily than at other times. A clear mucous ran from their noses, and explosive coughs interrupted the stillness of the forest. Respiratory troubles and other diseases are probably the principal causes of death in gorillas. It is well known that the animals easily die from pneumonia in captivity. Wild gorillas in West Africa seem to suffer from yaws, a disease which eats the face away, and from a leprosy-like infection. Skeletons in museums show that gorillas occasionally have arthritis. The Kabara gorillas seemed on the whole quite healthy. Many of the diseases of lower altitudes were not found at these heights: there were no mosquitoes, black flies, ticks, or other common carriers of parasites. When I put samples of dung under the microscope, I was surprised to find the eggs of a roundworm resembling the human hookworm in over half of the samples. Although a hookworm infestation may not in itself be serious, it can lower the resistance of an animal, especially when it is old, and make it susceptible to more serious disorders.

How do the gorillas respond to a member of their group which is sick or has just died? Do the animals "cover his corpse with a heap of leaves and loose earth collected and scraped up for the purpose" as suggested by the anatomist Owen? Do they mourn the death of a group member, a lifelong companion with whom they had daily social contact? Or do they simply abandon the corpse or sick animal to be ultimately devoured by scavengers and maggots? According to published accounts by hunters, the wounded and dead are usually left behind while the others flee. But the hunter Fred Merfield, stated that "gorillas never abandon their wounded until they are forced to do so, and I have often seen the Old Man trying to get a disabled member of his family away to safety." A most fascinating series of observations was made at Kisoro. In November, 1959, tourists noted that one member of a gorilla group suffered from diarrhea. On February 6, the sick animal, which happened to be the leading silverbacked male of the group, accompanied by a large infant, was encountered in the forest without sign of other members in the vicinity. These two animals, the large male and the infant, continued to roam together without returning to the group. They were seen on February 22, and on the following morning the male was found dead on the mountain side with the infant keeping its lone vigil beside him. When the Africans attempted to catch it, the youngster refused to leave the body of his dead companion. It was a brutal choice for such an infant to have to make: escape man and enter the forest to wander alone in search of its group, a task for which it was unprepared, or cling to this last vestige of its former happy group life, a dead leader who for the first time failed to protect it. Finally the youngster was captured, only to die later in the London zoo. A veterinary performed an autopsy on the male and found that his intestinal tract was highly inflamed and that his body was emaciated, due no doubt to constant diarrhea. The sick male had apparently abandoned his group to live in solitude until death overtook him. The infant had accompanied him perhaps inadvertently and certainly without realizing that it would not again see its mother or the rest of its group.

Since group IV, which I had watched most of September, had moved into the bamboo zone, I now searched the forest north of camp. Soon I found a new group, group VI, which consisted of one silverbacked male, one blackbacked male, nine females, two juveniles, and seven infants. The leader of the group was a large, rangy male, whom I named Dillon, because his long face bore a striking resemblance to a certain television entertainer. He was an unpredictable fellow whose moods changed from day to day. Sometimes he displayed a curiosity and forwardness matched by no other silverbacked male around Kabara, but on other days he roared at me and beat his chest, obviously not at all pleased at my presence. Later in the month a very distinctive and childless female disappeared from this group, and I never saw her again. Perhaps she died, or perhaps she simply switched her allegiance to another group. Another female, Brownie, was pregnant then, but I did not know it. She was a placid creature with sagging breasts and a chocolate-brown pelage. Her infant was born during the first two weeks of January, and she then became exceedingly shy as she scurried through the undergrowth, rarely letting me catch a glimpse of her baby. When the infant was one and a half months old she placed it on her back. This surprised me, for the youngster was obviously unable to hang on securely, and most mothers did not begin to transport their infants in this manner until they were three to four months old. I did not see the group again for a month, but in early April it returned to the vicinity of Kabara. Brownie no longer carried an infant. Some mishap had befallen it, and I wondered if perhaps the casual manner with which she had handled her offspring was the cause of its death or disappearance.

Between August, 1959, and August, 1960, a total of thirteen infants was born into the groups which I had under frequent observation. Of these one died, one was so seriously injured that it probably succumbed after I last saw it, and Brownie's baby disappeared and was undoubtedly dead. Thus, during this one year period, the mortality of infants during the first year of their life was 23 per cent. Nelson, a doctor at a mission near Kisoro, told me that the infants of Africans in the area probably have a death rate of similar magnitude. Based on various other computations, I estimated that between 40 and 50 per cent of the gorilla youngsters die before the age of six years. Gorillas, however, also have a high birth rate. Of twenty-seven females whose status was traced for over ten consecutive months, only two females lacked infants and failed to give birth to one, and both of these were elderly and physically below par: one had a cancerous-looking eye, the other a skin rash. Using the natality statistic births/1000 as applied to humans, the Kabara gorillas showed a yearly birth rate of 90/1000, which is about twice that of humans in countries with a so-called population explosion. Gorillas, of course, have no health services, and their death rate is correspondingly high.

The guard and I were hurrying home late one day after

having spent a few hours with group II. The forest was gloomy under its canopy of gray clouds, and even the songs of the sunbirds had a subdued quality. Suddenly, out of some wisps of fog, several gorillas climbed onto a log and sat down. I scanned their faces, noticing their noses, which in gorillas are highly distinctive, and realized that these were strangers to me, that yet another group used this part of the forest. I returned to this group (group VII) the following morning and settled down within forty yards of them. Eight juveniles, females, and a silverbacked male ascended two trees to a height of thirty-five feet and squatted on the branches, looking me over. I was immediately impressed with the placid and amiable countenance of the silverbacked male. The temperaments of gorillas vary as much as those of humans, and if the leading male of the group is irascible and excitable, I found it difficult to observe the animals. For example, I was never able to watch group XI for long because the male was so shy that he dashed away, followed by the other members of the group, whenever I appeared. But the male of group VII now sat peacefully in the crown of the tree, his large angular body seemingly out of place in this arboreal environment. This male had a predilection for climbing, and I dubbed him the Climber. He was an average-sized fellow with a rather harsh look about his face, for the lines of his mouth curved downward and the hairs on his chin were scraggly and unkempt. But the eyes beneath his beetling brows were unusually soft and warm. Once he shook a branch violently and once he beat his chest as if to intimidate me. Then he descended to a lower branch, beat his chest again, yawned, defecated, lost his balance and nearly fell, and finally climbed all the way down. In this group there were in addition to the silverbacked male, two blackbacked males, six females, four juveniles, and five infants. Most of these animals poked silently around in the undergrowth, snacking and just sitting. One juvenile squatted in the crotch of a *Hypericum* tree and broke off slivers of dry bark which it ate like a chocolate bar. A female ambled up to a juvenile, swiveled around, and shoved her rump almost into the face of the youngster. The juvenile then groomed the female, carefully parting and inspecting the hairs on her rump. At noon the animals moved slowly away, alternately walking and feeding, only the tops of their heads visible above the rank weeds.

As the months passed I became better acquainted with this group than with any other, for the central part of its range happened to fall near Kabara. Day after day I visited these animals until I felt that I knew and understood them, much as I would a human child before it is able to talk. Many scientists frown on the tendency to interpret the actions of animals in anthropomorphic terms, to read one's own feelings into the behavior of creatures, even if they are as closely related to man as the gorilla. But animals frequently do resemble man in their emotional and instinctive behavior, although, unlike man, they are perhaps not consciously aware of their own thought processes. I feel that something vital in our understanding of animals is lost if we fail to interpret their behavior in human terms, although it must be done cautiously. If a person thinks he understands a creature, he must be able to predict its behavior in any given situation, and with gorillas I was able to do this only if I followed the bare outline of my own feelings and mental processes. Only by looking at gorillas as living, feeling beings was I able to enter into the life of the group with comprehension, instead of remaining an ignorant spectator. Sir Julian Huxley expressed it in the following way in the *Journal of the Royal College of Surgeons of Edinburgh*:

It is also both scientifically legitimate and operationally necessary to ascribe mind, in the sense of subjective awareness, to higher animals. This is obvious as regards the anthropoid apes: they not only possess very similar bodies and sense-organs to ours, but also manifest similar behaviour, with a quite similar range of emotional expression, as anybody can see in the zoo; a range of curiosity, anger, alertness, affection, jealousy, fear, pain and pleasure. It is equally legitimate and necessary for other mammals, although the similarities are not so close. We just cannot really understand or properly interpret the behavior of elephants or dogs or cats or porpoises unless we do so to some extent in mental terms. This is not anthropomorphism: it is merely an extension of the principles of comparative study that have been so fruitful in comparative anatomy, comparative physiology, comparative cytology and other biological fields.

Between October, 1959, and September, 1960, I observed group VII for a total of 159 hours, sleeping with it during some nights and traveling with it during many days, becoming more and more involved and familiar with the daily life of the animals. It was a life of leisure, of eating and sleeping and ambling about, with little to break the rhythm of daylight and darkness. Yet they never seemed bored with this existence as many a man would probably have been with a similar routine. "One of the essentials of boredom," as Bertrand Russell has pointed out, "consists in the contrast between present circumstances and some other more agreeable circumstances which force themselves upon the imagination." Gorillas live in the present, taking life solely as it comes, and seem eminently contented with their lot. The essentials of their existence are there for the taking: food, nest materials, companionship, mates. There is no competition for these items, and what with everyone fully aware of his status in the group, there is rarely any strife beyond the occasional bickering which is apt to occur even among the most congenial companions. The animal psychologist H. Hediger once stated that man differs from other animals in not being chronically frightened. I have observed many kinds of animals and have yet to see a species which is habitually full of fear while pursuing its natural existence, except on those occasions when a predator is sensed. And I have never watched a more completely placid and relaxed animal than the gorilla. What does the gorilla have to fear? What animal would dare attack a gorilla, except man and very rarely the leopard? The gorilla's every action testifies to the fact that it feels itself the undisputed ruler of its mountain domain: it will flee from that which is new and strange and it will defend itself when threatened, but it never seems apprehensive about anything. Even such large and powerful creatures as the black buffalo were treated somewhat condescendingly. Once, for example, a bull buffalo aroused the interest of Mrs. Gnath, a rather large and forceful female of Group VII. The bull dozed in a lobelia patch, sporadically chewing its

cud. Mrs. Gnath saw the bull at seventy feet and walked directly
toward him without concern, her small infant riding on her back.
Only thirty feet separated the buffalo and the gorilla, with the
gorilla still advancing, when the bull panicked and, with tail held
stiffly horizontal, crashed away into the undergrowth.

The life of group VII was typical of the various groups around
Kabara, and the year, during which I followed the fortunes of
these animals, included many of the major and minor events that
a gorilla encounters in the course of a lifetime. Each day typically
began as the one of April 2. At half-past five in the morning, before the first birds had awakened, and while the forest still lay
muted under the fading stars, I eased out of my sleeping bag and
approached the sleeping group. When I was within fifty feet of the
nearest animal, I ascended a tree and settled myself in a broad
crotch. As it grew light, I could distinguish the various animals
in the nests. Mrs. Blacktop slept on her side, her small baby
enveloped in her mighty arms and cuddled close to her chest. Mrs.
Patch lay on her belly, arms and legs tucked under, and her baby
slept beside her. Somewhat off to one side and at the periphery of
the group was Mr. Shorthair, a blackbacked male, the bristly hair
on his crown standing very erect. His breathing rate was regular,
about fourteen inhalations a minute, as he slept blissfully into the
new day. Mrs. February too reclined on her side, her newborn
almost hidden in her arms. Only three feet from her huddled a
juvenile in its crude nest of lobelia stems. Several others, including the Climber, who was the leader of the group, were partially

or completely hidden in the undergrowth. At 5:40 the birds started to sing; at 6:00 it was light; at 6:40 the first pale rays of the sun passed over the reclining gorillas. But the animals slept on. Finally, at 6:52, Mrs. Blacktop sat up in her nest and rather morosely stared down at her paunch. Then she lay down again. At 7:15, Mrs. Patch woke and looked around with sleepy eyes. She stretched one arm far to the side and greeted the day with a cavernous yawn that exposed her black, tartar-covered teeth when she curled back her lips. Then she pulled in a piece of vine and peered as if entranced at this bit of food in her hand. Two other gorillas sat up in bed and vacantly gazed into space. One female reached over and broke off and ate a piece of celery: breakfast in bed. She squatted in her dung, which littered the nest cup and rim, but she paid no attention to her feces. Gorillas regularly foul their nests — often in prodigious amounts. Obviously gorillas are too lazy to get up at night and relieve themselves elsewhere, although sometimes the animal will first wriggle its bottom to the edge of the nest and then defecate. But why get up? The dung of gorillas is quite solid, not at all messy, and the hair of the apes is not soiled by the contact. Gorillas, like all monkeys and apes, usually defecate wherever they are. They move over the ground and through the trees, and they have no further contact with their own fecal matter. Only man, who occupies permanent abodes, has need to move to a certain place to relieve himself. In this he resembles many burrowing rodents which in their holes have certain spots reserved for defecation. To train monkeys, apes, and men to defecate at a certain place is not at all easy, suggesting that the behavior is not a part of their natural behavioral repertoire. By 7:30, three gorillas had left their fouled nests and were snacking nearby. The Climber rose at 7:45, and all members of the group immediately left their nests and crowded around him. The morning had begun in leisurely fashion, but now the first order of the day was to eat enough to fill each colossal belly.

There are many kinds of plants around Kabara, but gorillas ignore all but twenty-nine of them in their feeding. Their daily staples consist of bedstraw (*Galium*), wild celery, thistles, and nettles. When in season, bamboo shoots and the blue fruits of *Pygeum* may be taken. And when the mood strikes them, they may bite the bark off several trees or the base of a sedge leaf. But all in all, they are not catholic in their tastes, and I could not determine just why gorillas eat certain plants and ignore others. I found that, with few exceptions, the food plants tasted bitter or otherwise unpleasant to me.

I never saw gorillas eat animal matter in the wild — no bird's eggs, insects, mice, or other creatures — even though they had the opportunity to do so on occasion. Once a group passed over a dead duiker without handling the fresh remains, and another time a group nested beneath an olive pigeon nest without disturbing the single egg. In captivity, however, gorillas readily eat meat, and the pair at the Columbus zoo enjoyed a slab of boiled beef for dinner every day. In contrast, various observers have seen free-living baboons catch and eat the young of antelopes, and Jane Goodall watched chimpanzees in Tanganyika kill and devour monkeys.* The natives also told me that gorillas frequently raid the nests of wild bees, but around Kabara the only nest-robbers were the Batwa, which were so elusive that I only found their child-sized footprints and the smoking remnant of the tree which they had burned to get at the bees. Several gorillas once climbed up on a hollow log which contained bees, but the apes seemingly ignored the combs which were clearly visible.

After leaving the nest site, group VII spread out, each animal

* When I looked through the government files in the town of Kigoma, Tanganyika, I came on the following record of meat-eating by chimpanzees. In March, 1957, about five miles north of town, a woman walked along a path carrying her child on her back. The inquest report describes in the woman's words what happened next: "Then suddenly from the bush came a chimpanzee. We were in the bush and the village was far. I was tying up my *Kuni* [firewood]. I ran away and the chimpanzee hit me twice. He was about four feet tall; I fell down. Then it caught the child who was on my back. I made a great deal of noise and other women came. Then we saw the chimpanzee eating the child's ears, feet, hands, and head." The medical officer in Kigoma examined the body and found that the "scalp was removed and there were five depressed fractures of the skull. Both hands were missing. . . . One half of the right foot was missing. The injuries were consonant with a child's being eaten by some animal. The fractures of the skull were caused by teeth."

Junior struts along a log.

A typical foggy day at Kabara.

The dark summit pyramid of Mt. Mikeno and the lichen-festooned trees give the forest an otherworldly air.

Two white-necked ravens visit our meadow daily and feed from Kay's hand.

Kay prepares the evening meal on her tiny wood stove.

feeding silently, intensively, stuffing a wad of greenery into the mouth with one hand while reaching for more with the other. Characteristically, the gorilla sits and reaches for food in all directions, only to move a few steps and sit again. The only sounds are the snapping of branches, the smacking of lips, and an occasional belch. Infants stay with their mothers and probably learn what to eat and what not to eat by watching them. In this way, food habits are handed down from generation to generation, a primitive form of culture. Once I saw a youngster take a partially eaten *Vernonia* branch from the hands of its mother and gnaw out the remains of the pith. And on another occasion an infant pulled down the lip of its mother and extracted a bit of bed straw which it then ate. Sometimes infants bite into plants which the adults do not use. One tiny fellow stuffed some lichen into his mouth but hurriedly spat them out. When another infant began to gnaw on the petiole of a *Hagenia* leaf, which adults do not consume, its mother reached over and took the leaf away from it. Yet in all my hours of observation I never saw a mother actually hand food to her infant, like Kortlandt, the animal behaviorist, who watched female chimpanzees in the wild give pieces of paw-paw fruit to their young.

Infants also have to learn how to eat certain foods. The leaf of the bed straw, for example, has three rows of small hooks, which readily adhere to fur and are abrasive to the skin. Adults handle the vine carefully, as one entry from my field notes illustrates:

A sitting female reaches forward and with her right hand bends some *Senecio trichopterygius* toward her, and with her left hand pulls off strands of *Galium*. After examining the vine closely, she removes several dry leaves with her lips. Then she picks out several dry *Galium* stems between thumb and index finger, using first one hand and then the other. Finally she pushes the *Galium* several times against her partly-opened lips while twisting the vegetation around in her hand, thus forming a tight green wad in which all leaves adhere to each other. She stuffs the mass into her mouth and chews.

Small infants have not yet learned to make this tight wad, and they laboriously stuff the long strands into the mouth.

Feeding slows down considerably as the morning progresses,

and the apes become leisurely gourmets as they wander about, choosing a leaf here, a bit of bark there. The following observation from my report illustrates this leisurely feeding behavior in Junior, the blackbacked male of Group IV, during a half-hour period.

Junior sits and peers intently at the vegetation, reaches over, and bends the stalk of a *Senecio trichopterygius* to one side. He stretches far out and with a quick twist decapitates a *Helichrysum*. After stuffing the leafy top into his mouth, he looks around and spots two more plants of the same species, which he also eats in similar fashion. He then yanks some wild celery, including the root, from the ground, and with rapid sideways and backward jerks of the head bites apart the stalk before gnawing out the pitch. The sun appears briefly, and Junior rolls onto his back. But soon the sun hides behind a cloud, and Junior changes to his side, holding the sole of the right foot with the right hand. After about ten motionless minutes, he suddenly sits up, reaches far out, slides his hand up the stalk of a *Carduus afromontanus*, thus collecting the leaves in a bouquet which he pushes with petioles first into his mouth. This is followed by a leafy thistle top, prickles and all, and a *Helichrysum*. He then leaves his seat, ambles ten feet, and returns to his former place, carrying a thistle in one hand and a *Helichrysum* in the other. After eating the plants he sits hunched over for fifteen minutes. The rest of the group feeds slightly uphill, and Junior suddenly rises and moves toward the other members, plucking and eating a *Helichrysum* on the way. A giant senecio has been torn down by another gorilla, and Junior stops and rips off a leafy top. From the stem he bites large splinters until only a two-inch section of pith remains in his hand, which he eats. A strand of bed straw follows, and just before he moves out of sight, a final *Helichrysum*.

Gorillas require about two hours of feeding to become satiated in the morning, and each animal is so intent on filling its belly that it has little time for anything else. Scattered over fifty yards of terrain, the gorillas are frequently out of each other's sight in the dense undergrowth. But movement is usually so slow that there is little danger of their becoming separated from the group. Between nine and ten o'clock, the foraging generally comes slowly to a stop. The great amount of forage which the gorillas have by then consumed provides not only the needed nourishment but also the water. I have never seen gorillas drink in the wild. There is little permanent water at Kabara, for the runoff rapidly disappears in the porous soil and rock. But the vegetation is succulent, rain is frequent, and dew lies heavily on the foliage, all providing moisture for the gorilla's diet.

From midmorning to midafternoon is siesta time. The members of the group are pictures of utter contentment as they lie crowded around the silverbacked male, especially when the sun shines warmly on their bodies. Many loll on the ground, lying on their back, belly, or side, arms and legs askew; others lean in a sitting position against the trunks of trees. If gorillas had a religion, they would surely be sun worshippers. Once in a while a gorilla constructs a nest in which to rest. These nests are in no way distinguishable from those built for sleeping at night except that the packed-down vegetation of the night nests indicates that they have been occupied for a long time. On December 14, one of the juveniles in group VII built himself such a nest in a *Hypericum* tree some twenty feet above ground. The youngster stood in a crotch and pulled in small pliable branches with one hand and stuffed them down under its feet. After breaking in about ten branches in this manner, the ape constructed the basic platform of the nest. The juvenile then pushed down on the mass with its hands and feet. Next it pulled in all protruding twigs, broke them, and packed them down along the nest rim. As a final touch, it snapped off several small branches that still reared up from the nest. The construction time was five minutes.

In effect, the leading silverbacked males of groups are dictators who by virtue of their size and position always get their way. But these males also are tolerant and gentle, and this is especially evident during the periods of rest. The females and youngsters in the group genuinely seem to like their leader, not because he is dominant, but because they enjoy his company. Sometimes a female rested her head in his silver saddle or leaned heavily against his side. As many as five youngsters occasionally congregated by the male, sitting by his legs or in his lap, climbing up on his rump, and generally making a nuisance of themselves.

The male ignored them completely, unless their behavior became too uninhibited. Then a mere glance was sufficient to discipline them.

Sleeping, dozing, and sitting are not the only activities of the gorilla during the noon siesta. Some animals may be engaged with their morning toilet, scratching and grooming themselves. Much of the scratching starts in an exploratory manner with the fingers being merely moved against the grain of the hair, much as a man searches through the hair on his head. When something is noted, it is scratched off and occasionally eaten. A gorilla may groom itself very intently by slowly bending aside the hairs on the upper arm, or on another part of the body, to expose a piece of skin, which is then examined closely for possible imperfections, flaky bits of skin, or parasites. Females groom themselves twice as frequently as males, and juveniles do so more often than females. I could not find out why this difference exists, but somehow I cannot quite imagine that the average youngster merely shows a greater concern over his cleanliness than a female. Infants groom themselves rarely, probably because their mothers usually do it for them. When the youngster is small, the mother usually lays it in her lap or drapes it over one arm, then carefully grooms its pelage by parting the hairs. At such times her lips are pursed and her eyes watch her active fingers from a distance of six inches or less as if she were terribly short-sighted. She pays special attention to the cleanliness of the rectal region, which in youngsters is marked by a tuft of white hair like a fluff of cotton. The infants do not at all enjoy being turned upside down to have their anus inspected. They wiggle and squirm and kick, but the mother firmly persists with utter calmness. Never was a young gorilla chastised by being slapped. As the infants grow older, they tend to be groomed in a more cursory manner. For example, one two-year-old was sitting beside its mother when she reached over, gathered it into her arms, and affectionately nuzzled its shoulder before grooming its arm. The infant suddenly reached up and put both arms around her neck in an embrace.

It has frequently been asserted that, in general, grooming

strengthens and maintains the social bond between members of a monkey or ape group. Gorillas, however, have the disconcerting habit of refusing to comply with the premature generalizations that have been made about primates. If indeed mutual grooming strengthens the social bond in gorillas, one would expect the activity to be frequent between adults, as is the case in baboons. Yet I never saw a female groom a silverbacked male, and between females I saw it happen only five times. Once, for instance, two females sat near each other. One reached over, tapped the other on the upper arm with the back of the hand, rose, and pointedly turned her rump toward the face of the sitting female. The latter then groomed the indicated area. Thus, grooming between adult gorillas seems to be mainly utilitarian, confined primarily to the rump and back and other parts of the body which the animal itself cannot reach with ease.

Juveniles seemingly use grooming as a means of initiating social contact with a female. Juveniles are at an awkward social age, being too young to function as full members of the group, yet too old to receive the preferential treatment that is accorded infants: they are like teenagers in our society. Occasionally a juvenile ambles by a female, presumably its mother, and suddenly begins to groom her or her small infant, thereby achieving the desired social contact. When, on rare occasions, a female does not want the juvenile by her side, she merely snaps at or swats it. Juveniles sometimes take small infants gently out of the arms of their mothers and carry them a short distance. But when the infant squirms or in any way shows distress, the mother immediately rescues it.

For the youngsters the noon siesta is the only extended period of the day when they can play and wander about without danger of falling behind the rest of the group. Gorilla children, like their human counterparts, find it difficult to sit still, especially when mother obviously would like some peace and quiet. The spirit of adventure overtakes the infants when only three months old and barely able to crawl, and their mothers have to remain ever alert in order to protect them and to keep them within easy reach. As the young grow older, they are permitted to stray farther afield. By the age of eight months they may wander up to twenty feet from their mothers, and by one year they roam throughout the resting group, sometimes playing by themselves or with each other. The inherent reserve of gorillas even finds its expression in play. On the whole, young gorillas are not consistently playful, and occasionally days went by without my seeing a single instance of it, especially when the clouds hung low in the saddle and the vegetation was wet. I observed ninety-one instances of play involving one hundred and fifty-six animals, and of these all but five were infants and juveniles. Free-living gorillas seem to lose their playfulness by the time they are six years old, although once, when a female sat contentedly by a trail with her small infant, she reached over and slapped the rump of a blackbacked male. He in turn flicked her on the shoulder with his hand and ran a few steps. Two minutes later he dashed past her and pushed her lightly on the shoulder, a bantering gesture which she ignored.

I always enjoyed watching the youngsters play with uninhibited exuberance. About half of the youngsters played alone, and infants did so twice as often as juveniles. Most lone play consisted of swinging back and forth on lianas, of sliding down stems and slopes, and of waving arms and legs around with abandon. One rather demure little fellow placed the leafy head of a lobelia upside down on its head and sat motionless under its green hat, resembling a dark-visaged leprechaun. Another balanced a cushion of moss on its nape and gingerly walked back and forth, appearing quite proud of its adornment. Sometimes, overcome with *joi de vivre*, a sedate youngster broke into a gallop, threw itself on its side, rolled over, and continued on like a bouncing, furry ball. Adults, even though they often took the brunt of the youngster's exuberance, were highly tolerant. One eight-month-old, for instance, bumbled around by the resting silverbacked male. With a wide overhand motion it swatted the male on the nose, but he merely averted his head. The infant then ran downhill and turned a somersault over one shoulder and ended up on its back, kicking its legs in bicycle fashion and waving its arms above its head.

When playing together, the youngsters for the first time in their lives have the opportunity to come into close social contact with each other. Wrestling is a favorite pastime, and usually the arms and legs flail like windmills as the young roll over and over. Another frequent game is follow-the-leader, with the route going up trees, across fallen logs, down lianas, and perhaps across the belly of a dozing female. King-of-the-mountain is played on stumps and in bushes. As one youngster tries to storm the vantage point, the defender kicks the attacker in the face, steps on his hands, and pushes him down. Anything seems to be fair. Yet no one was ever hurt in such games, nor did they ever end in a quarrel. When large juveniles played with infants, the former merely contained their strength, as one brief sequence shows. A juvenile and an infant sat about four feet apart. Suddenly the juvenile twisted around and grabbed for the infant, which rushed away hotly pursued and obviously enjoying itself, for its mouth was open in a smile. Soon the juvenile overtook the infant and covered it with its body. Twisting and turning, struggling and kicking, more and more of the infant emerged from beneath the juvenile. Freedom gained, the infant grabbed a weed stalk at one end, and the juvenile snatched the other end. They pulled in opposite directions until the infant was yanked forward, no match for the juvenile. They then sat facing each other and grappled slowly. Another juvenile came dashing up and in passing swiped at the others, and all three raced away. Usually only two youngsters played together, but sometimes three or four did so for awhile. Once three juveniles left a rest area together. One grabbed the rump hairs of the first one with both hands, the third animal did the same to the second one, and then all three careered wildly down a slope in snake-dance fashion.

When by chance a game became too exuberant and rough, an infant showed this by crouching down submissively, arms and legs tucked under, presenting only its broad back to the opponent. An adult female receiving the worst of it in a quarrel with another female may employ the same submissive posture, and the other animal always respects the gesture and refrains from attacking further. This submissive gesture has a striking parallel in man when, for example, captives beg for mercy by crouching down or commoners kneel before a king. In fact, crouching has become ritualized in man, so to speak, and now functions also as a greeting. The deep bow of Orientals and the nodding of the head of Europeans when passing each other on the street are greetings, but both seem to be basically remnants of the submissive crouch, conveying that no aggression is intended.

When not playing, infants usually sit by and often slightly behind their mothers. In this way a youngster can simply reach out and grasp the long hairs on her back and be lifted automatically aboard when she decides to move to another spot. Juveniles sometimes like to roam around the resting group, snacking on a strand of bed straw, or climbing a tree to look out over the forest. Juveniles climb into trees about twice as frequently as females and infants, and four times as often as silverbacked males who, although capable climbers, seemingly have a disinclination to do so. On the whole, gorillas spend roughly 80 to 90 per cent of their waking hours on the ground, and when they do decide to climb, their actions are slow and deliberate. They climb up the trunk with the ability of a ten-year-old boy, grasping a branch and finding a foothold while advancing with the other arm and leg. Sometimes a support of dubious quality is first tested by jerking it before the animal trusts it with the whole weight of its body. Gorillas are not agile, and they are relatively poor judges of what branches can support their weight. Several times I have seen gorillas step on a twig which snapped, and the animal was only able to save itself from a serious tumble with a firm handhold. Once, in the distant past, gorillas were probably arboreal, but now with their huge size and swaying belly they seem anachronistic among the boughs. Descent from a tree is easily accomplished, feet first and facing the trunk. If there are no branches, the animals frequently use the soles of their feet as brakes and slide down hand over hand in a shower of bark.

Since the gorillas are so closely associated day and night, tempers naturally become a little frayed at times, usually for trivial reasons. Quarreling is usually confined to the females, with the silverbacked male listening in aloof silence to the annoyed barks, which are husky and short like that of dogs. But bickering sometimes turns into a free-for-all, with the females harshly screaming at each other in anger and grappling and biting. The male tolerates such a commotion only so long before he advances very purposefully on the females, emitting grunts. The screaming promptly subsides. No matter how violent a quarrel seemed, I never noted an injury as a result of it, for the animals always restrained themselves, and their biting was mostly in sham; they never fought seriously in my presence. A typical course of events during a quarrel is here related from my field notes:

A female walks leisurely past another one sitting by the trail. The latter slaps her on the back for a reason unknown to me. She in turn wheels around and runs with open mouth straight at the female who swatted her. This female cowers down with legs and arms tucked under, but with head raised screaming loudly. Her lips are curled up and the teeth and gums show. The two females then grapple briefly and mock-bite each other's shoulder. As the two fight, two other females run up and join the melee. All four then scream, grapple with each other, and run around with teeth bared. The rest of the group watches; that is, all but the silverbacked male who sits five feet from the nearest combatants and does not even turn his head. After about fifteen or twenty seconds, three females cease fighting and walk away. Only one female remains in the battle area and emits short screams. Suddenly she takes two steps after one of the retreating females and slaps at her hind leg; the one slapped turns and advances screaming. The first female backs away and collides with the silverbacked male, who gives an annoyed grunt. Two females meet and wrestle briefly as a third female runs up to join the hassle. Finally all part after the whole sequence has lasted about one minute.

The period of midday rest lasts anywhere from one to three hours. Sometimes, and for unknown reasons, the leader of the group decides to travel a few hundred feet, only to lie down and continue his nap, And, of course, all other members of the group follow him, for it is he who determines the basic daily routine by his actions — when and where to rest, how far to travel, where to go. The leader co-ordinates the behavior of the group with rather simple gestures and vocalizations. When, for example, he rises suddenly and walks with a rather stiff-legged gait in a certain direction, the others know not only that he is leaving but also the direction he is about to take. The females and youngsters then crowd around him in readiness to follow his bidding. Occasionally the leader stands motionless, feet spread, looking straight ahead. Again the others know that this posture heralds his imminent departure. The animals are frequently out of each other's sight in the dense vegetation, and the male, and occasionally a female, emits a very characteristic call, which seems to alert the group and to keep it together: a series of calls, rather low-pitched and abrupt, *u-u-u*. Sometimes, when the group was widely scattered I heard a curious call — a rapid, clear *ho-ho-ho* or *bu-bu-bu*, now and then rising and falling in pitch. Both these calls apparently mean "Here I am." The speed with which the gorillas respond to their leader depends to a large extent on his actions. There is obviously no need to hurry if he ambles away, but if he suddenly runs, danger might well be near, and the others race after him. In times of danger, any member of the group may assume temporary leadership. Once a juvenile saw me and with a high-pitched scream of fear bolted down the slope. All others in the group fled after this youngster even though they were unaware of the nature of the danger. It was a panic reaction which in man closely resembles the hysterical stampede which may follow the cry of "Fire!" in a crowded movie theater or restaurant.

As soon as the group begins to move out from the rest area, the infants dash to and climb on their mothers. Sometimes a female walks off before her offspring has reached her, and it then lets out a screech of distress which also carries quarrelsome overtones. The female may slow down and without turning around wait until her youngster has climbed up her leg and rump onto her back, or she may simply continue, blissfully ignoring her pursuing child. If there is danger, mothers naturally will dash to their young and gather them in. And once, when I suddenly met up with group IV, Big Daddy, the leader, grabbed a juvenile around the waist and holding it against his hip carried it to safety.

When moving in single file through the forest, the leader is often at the head of the procession, and a blackbacked male tends to bring up the rear. But as soon as the animals spread out to feed, obvious leadership ceases as each animal forages for itself yet keeps an eye on the behavior of others in the vicinity. Just how the leader decides where to go is a moot point, but often the group wanders so erratically that I suspect a lack of plan and purpose in his actions. In the beginning of the study I tried occasionally to outguess the gorillas by anticipating their direction of travel. More often than not I missed them because they had veered off to another part of the forest. Sometimes a group circled, or wound back and forth across its trail made earlier in the day; and at other times it advanced so slowly that the animals bedded down for the night only three or four hundred feet from where they had slept the night before. Occasionally they raced nonstop and very directly to a slope two miles away. On the whole, the only thing predictable about the daily movement of gorillas was its unpredictability. A group traveled on the average of about 1,100 to 2,100 feet a day, most of it in the afternoon between two and five o'clock. In some respects, then, gorillas can be considered nomads, moving along and feeding and finally bedding down wherever darkness overtakes them, only to begin another day of wandering.

But animals do not move at random through the world, and most cling to some piece of earth, which they know intimately. Throughout the study, I plotted the routes of travel of the various gorilla groups on a map, and soon it became apparent that there were some boundaries beyond which a group did not roam. In some areas, cultivation, grasslands, and other unfavorable habitats determined the boundaries of the gorilla's range. Around Kabara, however, the forest was relatively unbroken, and the animals seemed to have imaginary lines beyond which they did not venture. One such line ran from Rukumi to Kabara and up a prominent spur of Mt. Mikeno. Neither group IV nor group VIII crossed this boundary, and group VII skirted over it briefly only three times. It is probable that over the years each group becomes so familiar with its limited terrain that it hesitates to wander into new and strange areas. Infants learn to recognize the features of the boundaries past which the adult members of the group fail to wander with the result that, as the youngsters grow up, they too tend to remain within this culturally determined home range. Just how well gorillas know not only the grosser aspects of their environment but also the finer details was illustrated to me by group II. One ravine was a mere slash in the face of Mt. Mikeno, with perpendicular walls that dropped thirty feet in places. The easiest way to bridge this chasm was by way of a log that lay concealed behind some shrubbery. On two occasions, the gorillas walked directly to this natural bridge and crossed it, showing that they recalled its location.

A group ranged over a considerable area — from ten to fifteen square miles. Group VII, for instance, covered eight and a half square miles around Kabara, but the animals occasionally left the area for brief periods in the direction of Mt. Visoke. Group VI frequented eight square miles of forest, but it moved periodically into Rwanda, where I was forbidden to go. Within this range, groups followed an irregular pattern of movement, appearing in various parts at intermittent intervals. Group VI, for example, reached the slope behind Kabara about every forty days, with variations from fifty-two to seventy-eight days, and after staying there anywhere from two days to a month disappeared again in the direction from which it had come. Groups often had a temporary center of activity which was switched from time to time. Once, for instance, group VII remained for eighteen days in one and a half square miles of terrain near Kabara and then moved to another part of the forest. Gorilla ranges were not exclusively occupied by one group: as many as six groups visited the Bishitsi area off and on, and groups IV, VI, and VIII commonly used the slope of Mt. Mikeno near our camp.

By midafternoon the gorillas' day has consisted of sleeping, feeding, and sleeping, in that order, and the activity next in sequence is obviously feeding, which they continue until dusk. They forage leisurely, with much sitting around, alternated with occasional bursts of travel at two to three miles an hour. As the forest grows dusky, their movements become very languid, and they collect near the dominant male. If something, like my presence, disturbs them at this time, the animals immediately become restless and move away to bed down for the night far from any prying eyes. But if everything is peaceful, they sit around indecisively, seemingly waiting for someone else to make the first move. Then, about six o'clock, or as early as five o'clock if the sky is heavily overcast, the leader of the group begins to break branches to construct his nest, and the other members follow suit.

Ten to eleven hours after awakening, the animals are again ready to bed down after a strenuous day of feeding and resting. On April 1, I watched group VII as it traveled leisurely down a slope late in the afternoon. At 6:07 the silverbacked male, the Climber, was forty feet from his group and partially out of sight behind a low ridge. I heard the breaking of branches and knew that he was building his nest. Then a juvenile also began to construct its bed. It sat at the base of a tree and bent handfuls of small herbs toward its left side with the right hand. It then stood on two legs, grabbed the top of a mass of herbs, heavily overgrown with bed straw, and pulled it in. It sat and broke or bent the tips of the herbs to fit in a semicircle around its body before pressing the mass down with both hands. Standing on three legs, it reached far out and broke more weed stalks off at the base and pulled them in. After placing these along the edge of the nest, it broke their protruding tops to fit the rim. It sat, turned around, and sat again. The time required for building was about one minute. The other members continued to feed and move slowly until they were over forty feet away. The juvenile then abandoned its nest and joined the females who were sitting around in a lobelia patch, but the male remained where he was, quite separated from the others. On a few other occasions, several animals bedded down near the male while the rest walked on. As much as three hundred feet separated the two sleeping aggregations. At 6:30, the other members of group VII suddenly began to build nests, and five minutes later all was silent, except for one juvenile which continued to putter around for several minutes. Soon there was no sound in the forest at all, and as I unrolled my sleeping bag I found it difficult to realize that over twenty bulky gorillas slept almost beside me.

No aspect of gorilla behavior has received more study than that of nest building. Nests in trees and in bamboo may remain distinguishable for as long as one year after they have been built, and these structures are often the only evidence of gorillas encountered by a visitor to the animal's realm. Gorillas bed down where the night overtakes them, and the only requisite of a nest site is the presence of some vegetation with which to construct the nest. The apes build nests either on the ground or in trees, but the percentage of ground nests to tree nests varies from area to area. In the *Hagenia* woodland around Kabara, 97 per cent of the nests were on the ground because the branches of most trees were unsuitable for building. The foliage of the *Hagenia* clusters in the crown, and the twigs snap off when broken, making it difficult to construct a stable platform. Juveniles nested in trees twice as often as females and blackbacked males. Silverbacked males always slept on the ground at Kabara, but in nearby Kisoro a few occasionally built a nest in the shrubbery up to eight feet above ground. It is commonly believed that females and youngsters nest in trees for protection while the male crouches at the base to guard them. This is a myth. Some observers have also claimed that gorillas nest in a specific relation to each other and always within sight of a neighbor. Actually each adult builds its nest in a patch of suitable vegetation, usually in the vicinity of the leader but not necessarily within his sight or that of anyone else; blackbacked males quite commonly nest sixty feet or more from the rest of the group. Juveniles tend to sleep near females and in some instances in the same nest with them. A baby remains with its mother, except now and then when a large infant builds its own miniature nest beside her.

Infants begin to construct nests at an early age. One youngster, only eight months old, sat propped against the back of its mother. Twice it bent in a weed with one hand, then released it. Finally it pulled in a third stalk and pushed it under one foot, exhibiting in rudimentary form the typical behavior of adults. By the age of fifteen months, the young gorillas occasionally build practice nests in the evening, but these are rarely used for sleeping. Thus, young gorillas begin to construct nests over eighteen months before they are called on to use the behavior regularly in their daily life as juveniles. It is sometimes thought that nest building is completely learned by infants, but I doubt if this is true. The infant certainly learns how to construct a stable and secure nest in various locations and how best to use the different kinds of vegetation. But I believe that the gorilla has an inborn tendency to build something under and around its body before lying down to rest, a supposition which only adequate research on captive animals can prove. One observation made by Dr. E. Lang of the Basel zoo tends to support this idea. Goma, a ten-month-old female gorilla born at the zoo and raised in a human home, once placed branches around herself at dusk to form the crude outline of a nest.

Nest construction is a simple process, rarely requiring more than three to five minutes, and sometimes as little as thirty seconds. About 10 per cent of the animals at Kabara built no nest at all, and many ground nests consisted only of three or four handfuls of weeds packed down around the animal to form a partial rim. When building a ground nest, the gorilla merely bends and breaks in weeds, twigs, and vines from all directions and places the vegetation around its body without particular order. There is no interlacing, weaving, knot-tying, or other involved manipulation. Nests in trees have to be more substantial than those on the ground, for if they were to collapse the resulting tumble could well be dangerous to the animal. Tree nests tend to be built in forks and along horizontal limbs for support, and the surrounding branches are bent and snapped inward until a fairly stable platform has been created.

When I saw the crude gorilla nests on the ground, I often wondered what possible function they could have. They did not provide shelter unless placed against the leaning bole of a tree or in a bower of vines, and they did not offer insulation from the cold earth, for the emphasis of construction was on the rim rather than on the bottom. On steep slopes, ground nests often prevented the animal from sliding downhill, but I found that nests in such locations regularly disintegrated during the night, and the animal picked itself up in the morning ten feet or so below where it had gone to bed the previous evening. Ground nests seem to have little or no function. Nests above ground, on the other hand, have a very definite function: they provide a secure platform on which the animal can recline while sleeping during the night without danger of falling. It is, therefore, probable that the nest building habit developed in trees, when the gorillas' ancestor was still arboreal.

The gorilla is not the only ape which builds a nest — the chimpanzee and the orangutan do so too. When I examined chimpanzee nests in various Uganda forests, I noted that they were made like those of the gorilla. I observed chimpanzee nests only in trees, usually from fifteen to one hundred feet above ground, but Dr. V. Reynolds, who studied chimpanzees in the Budongo Forest of Uganda, told me that on two occasions he found a ground nest made by this ape. In November and December of 1960 Kay and I visited Sarawak in northwestern Borneo to make a brief survey of the orangutan. I found over two hundred orangutan nests in trees, between eleven and cne hundred and thirty feet above ground. The nests of the orangutan were just like those of the gorilla, except that many contained a thick lining of twigs which the ape had broken off around the nest, apparently to make its bed more comfortable.

One day, in Borneo, while the temperature was over 80°F. and the humidity stood at 100 per cent, Gaun anak Sureng, my Dayak guide, and I sloshed through the swamp forest that grows in the low country bordering the South China Sea. The water was ankle to chest deep in the forest, and every few minutes we halted to pluck the myriads of blood-sucking leeches from our skin. We were looking for orangutans but saw only old nests. I was just about to record the height of two nests in my notebook when Gaun shook his head and said "Bear." At first I did not understand what he meant, but when he indicated that bears had constructed the nests, I was dubious. Then he showed me the claw marks on the tree trunk which his discerning eyes had immediately noted. I shinnyed up the trunk to a height of forty-five feet, following the claw marks in the bark, until I reached the nests. I examined these nests closely and found that they were similar to those of apes. One nest consisted of four branches broken inward to form a platform, the other of six branches. The bark on each branch was marred by claw marks, and I felt certain that Gaun was right, that the bears had indeed built these nests with their paws. I had not realized that the Malayan black bear, a small, bowlegged creature with a blaze of white on its chest, constructed nests. When I checked previous writings for nest building in Malayan bears, I came across only vague references, like the one by E. Banks, who wrote in *Bornean Mammals*: "Said to make a nest of sticks in trees but this is not yet confirmed." Although I certainly would like further data, I find it interesting that a behavior pattern which is so often thought to be unique to the apes is possibly also shared by the bear.

The gorilla's life consists of sleeping and feeding and sleeping some more. Only at irregular intervals does anything break the routine of its existence. I no longer interrupted the activity of several groups; I was nearly as much a part of their natural surroundings as the duiker and the buffalo. But to those groups which I had rarely visited, I continued to be an object of great curiosity. Although the animals were somewhat afraid at times, they seemed to welcome the opportunity to inspect a new and strange animal. A meeting between two groups afforded the animals a break in their routine. Since as many as six groups used the same part of the forest at irregular intervals, it was only a matter of time until two of them met face to face. And it was only a matter of luck that I happened to be there to watch such a meeting. On the whole, a group tended to ignore another group if the two heard but could not see each other. Group VII, for instance, heard members of another group beat their chests or scream in the distance at least three times, yet it continued its activity without interruption. However, on four occasions, group VII established contact with a neighbor. In the early morning of

October 19, I saw group III — a male, two females, one juvenile, and an infant — near the slopes of Mt. Mikeno. Later in the day I discovered group VII resting quietly within two hundred and fifty yards of where I had last seen group III. Toward evening, after I had gone home, the two groups joined, quite amiably it seemed, and nested together for the night. At ten the following morning, when I reached the combined nest site, the groups had already parted, each going its own way.

About two weeks later, on November 1, group VII was in the saddle area near Bishitsi, where group V was often seen. During the night of October 31, the two groups slept about two hundred yards apart, and when I came upon them in the morning they were feeding and resting, each blithely ignoring the other with an indifference so studied that their interest must have been very great. After noon, a blackbacked male and three juveniles of group V filed up on a fallen tree and watched the activities of group VII. A juvenile of group VII approached within fifty yards of the quartet on the log and looked it over with obvious interest. The other gorillas continued to feed, and only the silverbacked males jerked up their heads to glance around every so often. By midafternoon interest in each other had almost completely waned; only the Climber of group VII beat his chest now and then, showing his excitement, and a blackbacked male of group V rested on a log, head propped in his hands, gazing at his transient neighbors. By four o'clock the groups were moving in opposite directions, and on subsequent days they continued their self-contained lives without further meeting.

Group VII met group XI on April 18. The Climber sat hunched over, staring at the ground in front of him, seemingly in deep thought, with his group clustered tightly around him The silverbacked male of group XI squatted only twenty feet away by a tree, with the fifteen members of his group scattered through the underbrush behind him. This male was excited, and he put on a spectacular show: he hooted softly and with increasing tempo until the sound slurred into a harsh growl; he beat his chest, wheeled about, lumbered up a log, and with a forward lunge jumped down to land with a crash. As a finale he gave the ground a hollow thump with the palm of his hand. The Climber walked rapidly toward the other male, and the two stared into each other's eyes, their faces a foot apart. These giants of the forest, each with the strength of several men, were settling their differences, whatever they were, not by fighting but rather by trying to stare each other down. They stared at each other threateningly for from twenty to thirty seconds, but neither gave in, and they parted. The Climber returned to his seat, but in the ensuing minutes he twice tried to get the upper hand without succeeding. The males reminded me of two belligerent schoolboys who are trying to intimidate each other without direct combat. The Climber made a final effort to win this war of nerves by throwing a handful of weeds into the air, as he stood bolt upright to reveal his full power, and suddenly racing at the other male to stare into his face from a distance of about an inch. But this male was as determined as he was excitable, and the bluff failed. The other members of the group paid little attention to the leaders; they acted as if they had no interest in the outcome of the struggle. The Climber began to feed after his last bid for supremacy, and his females and youngsters joined him. A blackbacked male and a juvenile of group VII passed the male of group XI and mingled with his females briefly and peacefully before returning to their own side. The Climber sauntered up to his opponent a final time and stared at him in a half-hearted way before wheeling around and walking rapidly away, followed by his group. The animals passed directly under the tree in which I sat. Although they were aware of me, they ignored my presence — more interesting things were happening today.

The mind of a gorilla has its own mysterious paths which even the most persistent observer may find difficult to trace. I expected the groups to part now, but for some reason the male of group XI began to follow the retreating animals alone. Twice the Climber dashed back over his trail, hurling his bulk at his persistent opponent only to brake to a stop just before physical contact. Unfortunately, group XI then spotted me and was distracted. I left soon after and the following morning read a strange tale on the forest floor. Group VII had apparently moved steadily ahead in single or double file, snatching a bite to eat in passing. And somewhere behind them followed group XI over the same trail. Group VII had just built its nests at dusk when the other group caught up with it. Group VII promptly abandoned its nests, moved two hundred feet, and made new ones. But the male in group XI was persistent, and he followed. Even the towering patience of a gorilla must have been worn thin by this incessant pursuit, for on the trail I found several tufts of hair pulled out by the roots. A tussle probably occurred. Group VII made a last effort to leave the other animals behind. It traveled in single file some two hundred yards and built very crude nests. Group XI, finally outdistanced, lay down on the trail to sleep and on the following morning moved back in the direction from which it had come. None of the animals in either group were the worse for wear, but I suspected that the Climber would think twice before again being neighborly with group XI.

The most striking aspect of these meetings was the highly variable response which group VII showed to the other groups. With group III it peacefully shared a common nest site; it approached group V to within fifty yards but made no attempt to mingle; and it behaved antagonistically toward group XI. Some members of group VII have undoubtedly roamed the saddle area together for fifteen years or more. In the course of their restricted wanderings, they probably have met the various neighboring groups many times, and they may have made friends and they may have made enemies. But, of course, I had no way of knowing of these past events.

Gorillas have very strong attachments to members of their own group, probably because they feel more secure and content among intimate friends and relations than among more casual acquaintances. When groups meet, mingle, and subsequently part, each animal tends to remain with its respective group. Rare exceptions to this occur. Between mid-February and mid-March a female with an infant joined group VII. Since I had neither seen nor heard of a lone female wandering around in the forest, I suspected that she switched her allegiance from one group to another when the two mingled or were side by side. I had never seen her before, and until she learned to know me, she created considerable confusion in the group. At the first sight of me, she tended to scream and dash off, much to the puzzlement of the other group members who had long ago ceased to regard me as an object worthy of note. The new female, I am sure, felt just as baffled about the behavior of her adopted group, which haughtily ignored her warnings in the face of man, the gorilla's worst enemy.

I never observed fighting between individuals of different groups, and the bluff charges and staring that occurred between the Climber and the male of group XI were unique in my experience. Gorillas are eminently gentle and amiable creatures, and the dictum of peaceful coexistence is their way of life. In this man would do well to learn from the gorilla. Of course, no animal reaches an absolute standard of perfection, and definite fights between males have been reliably reported. In July, 1958, two silverbacked males, each the leader of a group, fought several times for unknown reasons, according to Baumgartel, who obtained the information from his African trackers. On August 11 the males battled again, and on August 13 one of the males was found dead. The only evidences of fighting found on his body were a bruised right eye and bites on the backs of both hands. The cause of his death could not be determined by the doctors who examined the body at the medical school in Kampala.

The fact that several gorilla groups occupy the same section of the forest, and that, when groups meet, their interactions tend to be peaceful was of considerable interest to me. Once it was generally thought that each monkey and ape group lived in a territory, the boundaries of which were defended vigorously against intrusion by other members of the same species. But the gorilla certainly shares its range and its abundant food resources with others of its kind, disdaining all claims to a plot of land of its own.

Although group VII met other groups on occasion, it was never joined by a lone male. I identified seven lone males around Kabara — four silverbacked and three blackbacked ones — and there may well have been a few more. But these males associated solely with groups IV and VI, not with the others. Apparently only certain groups will accept lone males, and the loners have learned where they are welcome and where not. Of the lone males, I knew the Lone Stranger best. He was a silverback in the prime of life with rather brooding eyes and a supercilious slant to his mouth. He did not like the sight of me and usually bolted with a brief roar as soon as we met. I first saw him on November 18 at the periphery of group VI. Mr. Dillon, the leader of the group, rested surrounded by his females and youngsters, all seemingly quite oblivious to the Lone Stranger only thirty feet away. Mr. Dillon rose fifteen minutes later, and his group moved past him while he remained behind to stare at the Lone Stranger. Apparently Mr. Dillon was hinting, not at all subtly, that it was time for the lone one to be on his way. The Lone Stranger left the vicinity of the group that day, but in the following weeks I flushed him from the fresh trail of this group several times. When group IV moved onto the slopes of Mt. Mikeno in May, the Lone Stranger joined it for at least a week. But the wanderlust overtook him again, and he left to continue his lone life in the forest.

Two other lone males, a large blackback and an old silverback who appeared to travel together, were received more cordially than the Lone Stranger by group VI. On January 9 I saw the Kicker, a small blackbacked male of group VI, together with the two strange males about one hundred feet from the main group. A few minutes later the two blackbacked males joined the others, the stranger being accepted without response, but the new silverbacked male kept his distance. Dense fog rolled in, which happens at the most importune times, making further observations impossible. But on the following day, when I found the group, the new silverbacked male had joined the other members and now lay at ease on a hummock surrounded by several females. Mr. Dillon ignored the newcomers, but for some reason they had both left the following morning, and I never saw them again.

The comings and goings were even more pronounced in group

IV. Early in my study the Outsider parted and joined at least twice, and the Newcomer attached himself to the group. In January, after a three-month absence, the group reappeared in its old haunts. There were some drastic changes. Big Daddy was still the boss, but D. J., the number two male, was gone, as was the Outsider. Instead, there was a new silverbacked male who had taken over D. J.'s position. This new male had apparently brought two females and two infants with him, raising the total number of animals in the group to thirty. On April 24, Splitnose, who had been a member of the group at least since August, suddenly left, and I missed my vociferous friend. At least seven different silverbacked males associated with group IV, in the course of twelve months, but only Big Daddy remained the entire year. Big Daddy was the epitome of the tolerant and gentle gorilla, and I suspect that his permissive character had much to do with the casual visits of males to his group. Other males, like the Climber in group VII, were probably more possessive and discouraged potential visitors.

On the whole, the membership of groups remained stable over long periods of time. Births and deaths and the restless wanderings of lone males accounted for most of the changes. Nothing at all happened in group V over a ten-month period, or in group VIII over a seven-month period. At Kisoro there was a small group, consisting of one silverbacked male, one blackbacked male, three females, and one infant, which was seen many times by tourists between late 1957 and early 1960. In February, 1959, an infant was born, the only change in the group over a period of two years. But this little band fell on hard times. A year later, the silverbacked male, followed by a large infant, left the group. The male died of intestinal troubles, and the infant was captured and sent to a zoo. The survivors moved to Rwanda and apparently found a new leader — a silverbacked male who had only two females in his group. The combined group of eight animals had about one year of peace, and an infant was born during this period. But in February, 1961, the females were again deprived of their leader, this time by a black leopard turned gorilla killer. A female also fell victim to this predator. Once more the survivors went off in search of a new leader and protector, and, according to Reuben and his trackers, they apparently found one.

At this point it is perhaps useful to summarize and compare briefly what is known about the group organization of the other apes — the gibbon, orangutan, and chimpanzee — in order to put the gorilla into proper perspective.

The agile gibbon is the smallest of the apes, weighing from twelve to twenty pounds, but it is the most abundant, being found throughout southeast Asia from Assam and Burma to Thailand, Malaya, Sumatra, and Borneo. There are several species of gibbon, anywhere from five to twelve depending on the authority, but their habitat is similar, a tropical rain forest below an altitude of seven thousand feet. Only one species, *Hylobates lar*, has been studied. Dr. C. R. Carpenter observed twenty-one groups over a three-month period in Thailand and found that each group consisted of two to six animals. "The mean grouping tendency in the gibbons of the species *Hylobates lar* is that of the family; a male and female with their young." Each group ranges through the forest canopy as a cohesive unit covering an area of from thirty to one hundred acres, and feeding primarily on fruits, but also on leaves, insects, and nestling birds. Although adjacent groups may mingle briefly, they tend to stay in their own exclusive areas.

The orangutan (*Pongo pygmaeus*) is found only in certain parts of Borneo and northern Sumatra, where it is threatened with extinction because of indiscriminate hunting. This fruit-eating dweller of the rain forest canopy is bulky, with females averaging about eighty pounds in weight and males roughly twice that much. No detailed studies of the orangutan have been made, but my brief observations in Sarawak showed that groups are very small, consisting of only two to five animals each. These groups are unstable, with members apparently parting and joining. I saw a female with infant alone in a section of the forest, and another time I came across two sub-adults. During one night a male, a female with infant, and a juvenile slept in adjoining trees; on another night the male slept four hundred feet from the rest of the group; and on the third night the male had left the group.

Lone males, which many observers have encountered, undoubtedly join the females at irregular intervals.

The chimpanzee occurs throughout the rain forest belt from west to central Africa, and has even penetrated open dry woodlands in places like Tanganyika. Of the two kinds of chimpanzee, only the common chimpanzee has been studied; the pygmy chimpanzee, which inhabits the remote forests south of the Congo River, remains almost unknown. Vernon Reynolds, who worked for eight months in the Budongo Forest of Uganda, discovered that a group of from sixty-five to seventy-five chimpanzees inhabits roughly six to eight square miles of forest, but that, unlike the gorilla, the group is a loose and highly unstable aggregation of animals. Changes occur daily as chimpanzees stray singly and together in search of fruits, which form their principal diet in this forest. Sometimes several males roam together, and several mothers with infants are frequently found by themselves. Interactions with neighbors occur regularly, and there is much casual and peaceful mixing.

This brief summary is sufficient to show the fatuousness of making sweeping generalizations about the behavior of apes. In fact, striking differences may even occur within the same species in different parts of its range. For example, Reynolds never saw chimpanzees in the Budongo Forest using tools or eating meat, whereas Jane Goodall in Tanganyika observed both types of behavior.

Gorilla-watching as an occupation soon becomes routine, but I never found it dull and I never grew bored. Each walk into the forest filled me with anticipation, for the most interesting aspects of gorilla behavior, like the joining of groups, occurred at the most unexpected times. And whenever I thought that I must surely now know all groups in the area, a new one suddenly appeared. Group VIII, for example, turned up near Kabara in the middle of November, a full three months after I had started the study. It was a large group, consisting of one silverbacked male, two blackbacked males, eight females, three juveniles, and seven infants. Mr. Crest, the leader, had a huge sagittal crest; by

temperament he was excitable, yet bold. One particular morning with this group remains vividly in my mind. As I climbed onto a horizontal branch about five feet from the ground to watch the animals resting and snacking one hundred feet away, a female spotted me and beat her chest and then lay down. Happily, no one else appeared to note my presence, not even Mr. Crest, who in my previous eleven meetings with the group was usually the first to roar at my approach. One of the blackbacked males suddenly saw me an hour and a half later, and when he beat his chest, Mr. Crest jerked up his head and roared. The others in the group collected around him, and then all advanced very purposefully in my direction. To my relief they halted at thirty feet and looked me over intently. A female, with a three-month-old clutched to her chest, angled closer, peering at me out of the corner of her eye. When only fifteen feet separated the two of us, she reached up and gave the branch on which I was sitting a sharp jerk. She then glanced at me as if to judge the result of her effort, and finally pulled herself up onto the branch with me. Both of us squatted on the limb, casting fleeting looks at each other like a pair of strangers on a park bench. Having satisfied her curiosity, she swung down, and her place was briefly taken by a juvenile, who between glances in my direction bit off pieces of bark, seemingly out of nervousness. The final member to ascend the branch was another female. Interest in me then waned, and the group slowly retreated to its former rest area, where it continued its daily routine as if I did not exist.

The most striking female in this group was Olive Twist, so named for her grotesquely twisted face. Somehow, perhaps in a quarrel or during a fall from a tree, she had received so powerful a blow against the jaw that the whole lower part of her face was pushed sideways and her mouth was almost vertical. Although her jaw was probably broken, she showed no discomfort and she ably cared for her small infant. She was the only deformed gorilla among the animals I studied. In West Africa, the hunter Fred Merfield once saw a large male, the leader of a group, with only one arm, and he also encountered a lone male with a deformed foot. One anthropologist noted that in museum collections 2 per cent of gorilla females and 6 per cent of males had healed fractures of the skull. Very striking is the frequency of healed bone fractures in the arboreal orangutan. The anatomist Schultz found that 34 per cent of a group of one hundred adult specimens of the orangutan had healed fractures of the fingers, arms, and other parts. If the bulky orangutan, who is eminently adapted to life in trees, breaks its bones in mishaps this often, it is probably just as well that the still more ponderous gorilla tends to stick close to the ground.

Mrs. V., in group VIII, immediately attracted my attention because she seemed to have two infants. She carried one tiny young gorilla, about four months old, to her chest, and another, about a year and a half, toddled at her heels, often holding on to her rump hairs with one or both hands. It was not unusual for youngsters to sit by or even be fleetingly held by a strange female, but such young usually returned to their own mothers within an hour. At first I thought that perhaps Mrs. V. had borne young only one year apart, but after watching for several days, I noted that when the group traveled rapidly or when evening was approaching the large infant returned to a certain female, most certainly its mother. Why this infant preferred the company of Mrs. V., I could not fathom. Mrs. V. was a rangy animal with a bulging belly that looked ludicrous, and her face had a gaunt, ascetic mien. She was an efficient mother, with a certain unhesitating sureness in the handling of her own infant that bespoke of experience in baby care. She usually ignored the large infant, although she groomed it on occasion, yet the association persisted at least from November through May. The real mother of the infant never showed the slightest overt interest in the periodic absence of her young and this is perhaps a clue to the tenuous relationship between the two.

Group VIII, together with group VI, also furnished me with my most exciting three days of gorilla tracking at Kabara. At 9:45 on the morning of November 22 I came upon group VIII feeding slowly. Mr. Crest was staring uphill, and, when I followed his gaze, I saw Mr. Dillon, the leader of group VI, three hundred feet away. Slowly, almost imperceptibly, group VIII advanced in the direction of the other group. At 10:30, when only one hundred feet separated them, Mr. Dillon rose, took a few abrupt steps downhill, and squatted on a hummock, only to beat his chest and retreat uphill a few minutes later. Then, without warning, he walked rapidly at an angle toward Mr. Crest. The two huge males then sat thirty-five feet apart, backs toward each other, the most flagrant case of studied indifference I had ever seen. A little after 11:00, several members of group VIII wandered about one hundred feet up the slope to rest, and they were soon joined by Mr. Crest. For two hours the two groups held their siesta as if the other did not exist, and finally, after a few chest beats, group VI moved off in one direction without a backward glance, and group VIII soon retreated in another. I thought that their association had now ended, and what with the fog settling in on the forest, I crawled out from behind the log that was my hiding place and went home. But later in the afternoon the two groups traveled parallel to each other and finally bedded down side by side with only thirty feet separating the nearest nests of the respective groups.

When I found the groups again at 9:45, they were forty yards apart, with group VIII out of sight in a shallow ravine. Again the groups seemed unaware of each other until Mr. Dillon approached group VIII to within twenty yards. Slowly, Mr. Crest and his group advanced. Mr. Dillon retreated forty feet uphill, obviously uneasy, but his females and youngsters remained behind, and not until Mr. Crest was within fifteen feet of them did they withdraw. The animals then foraged actively and by 2:20 had moved out of each other's sight. There was no further contact that day, and I felt sure that they had seen the last of each other. I wondered why they had not mingled, when they seemed so calm about the whole meeting. But gorillas, I knew, are introverts, who keep their emotions suppressed. They retain their outward dignity when in fact they may inwardly be seething with excitement and turmoil.

On the third day, on November 24, group VI, which had so far taken all the initiative in maintaining contact, traveled rapidly some eight hundred feet and approached group VIII. At 9:30, the two groups moved out of a deep ravine side by side gradually mixing, until the two silverbacked males were only ten feet apart. Soon the leaders drifted in opposite directions, taking their own members with them. The two groups then occupied adjoining rest areas for three hours. Two juveniles and two infants from group VIII visited the edge of group VI and sat there side by side. Immediately two infants and a juvenile of the latter group hurried uphill to inspect the strangers from a distance of six feet. Like human children probing at and testing a new boy in the neighborhood, the gorilla young of group VI tried to goad the others into action. One juvenile jerked forward as if to charge, but its opponent merely turned and fled several steps. Then an infant sidled closer and with a wild lunge took a swipe at the leg of the other juvenile stranger. But it missed and dashed back to its own side. Dense fog rolled in, and when it lifted twenty minutes later the young of group VI were with their mothers. Soon only the two juveniles of group VIII were lying ignored at the periphery of the other group. A blackbacked male of their group then joined them. At one o'clock, when Mr. Crest rose and began to move away, he was as always followed by the females and youngsters. The blackbacked male and one juvenile tarried awhile near group VI until they too were sped on their way by a female who made a motion as if to charge. This time the two groups parted, leaving me to wonder why for three days they had remained so closely associated.

Another exciting period of tracking was provided by group IV. I came upon this group on April 25 as it was feeding near the rim of Kanyamagufa Canyon. The chasm lay between us, but the opposite slope slanted so steeply upward that I could see each animal. I immediately recognized Junior and several others, but to my surprise Big Daddy and many of the females were not in view. Then I happened to glance toward the lowlands, and there, six hundred feet away, was the rest of the group. Such a splitting of groups was very rare, and I wondered what had brought it about. Mrs. Wrinkle, who had been partially hidden behind some shrubs, ambled into view. She was a large, wooly-looking female with rather sad eyes and a face as wrinkled as a prune. The juvenile which usually tagged along behind her was not at her heels. Instead, she held a newborn gorilla, still greyish and wet, in her arms. The two units of the group united the same evening and nested together. When I again saw Mrs. Wrinkle the following day, she sat at the base of a tree in a shaft of sun, her infant craddled in her arms. She supported its head and limbs carefully, and for minutes on end gazed down at her feeble offspring with seeming devotion. I did not watch the group on April 26, but on April 27 I eagerly looked for Mrs. Wrinkle and found her with several other females. Her infant lay loosely in her lap. Her broad hand supported its back, but its arms and legs dangled loosely, its mouth was open, and its eyes were closed. It was dead. With the death of the infant, the behavior of Mrs. Wrinkle changed. She no longer held its tiny body close to her chest, nor did she bother to look at it. When walking she merely held it in one hand and even took several steps dragging it on the ground. The next day the crumpled body lay at her feet for two hours without being touched by her. Once she lifted the corpse by one hand, then dropped it. Yet she retained her young on this and the following day, and I only wished that soon she would abandon it. During the afternoon of May 1, four days after the death of the infant, she left it on the trail, face down, arms and legs askew. Flies buzzed around its putrefying body. I carried it home and measured it, with Kay taking notes for me at a distance to escape the horrible odor. The infant was nineteen inches long. As I stripped the soft meat from the fragile bones in order to preserve the skeleton, I felt sad, for it was like dissecting a human baby. At no time in their lives are apes and man more alike in appearance than in early infancy. The toothless gums and the bulging rounded forehead surmounting the rather flat face of the gorilla baby bear a striking resemblance to a newborn human. There is relatively little difference in the size of the brain case. The human infant at birth has a brain capacity of only 350 cubic centimeters, but by the age of one year the size of the brain has trebled, a phenomenal growth which is uniquely human. Even an adult male gorilla has a brain capacity of only about 600 to 700 cc.

Unfortunately, I never managed to see a gorilla birth or get much information on the course of pregnancy, data most easily obtained in zoos. No zoo bred gorillas before 1956, a sad record considering the large number of these apes that over the years have been held in captivity. The treatment of apes in zoos has been and often still is scandalous. The creatures sit alone behind bars, like prisoners in solitary confinement. Even today infant gorillas and other apes continue to be purchased at exorbitant prices only to pine away for lack of proper attention. I wonder why, for example, a zoo will pay $5,000 for a mountain gorilla infant and then be unwilling to hire a full-time nurse to provide it with needed companionship and constant care. A mild virus attack, which barely affects man, may be fatal to a young ape within a matter of hours if no help is available. Nor have I been able to see the logic of providing lions and other cats, which are content to do nothing at all, with sumptuous enclosures and then placing alert, active, and highly social apes alone in small pens. Young apes need the same kind and amount of attention given human infants. Zoos which cannot or will not provide their charges with such care should not be permitted to keep them, especially now that some species like the orangutan are in danger of extinction. If nothing else, man should show some ethical and moral responsibility toward creatures which resemble him so closely in body and mind. But then man has never learned to treat even his own kind with compassion.

As recently as 1915, William Hornaday, a prominent American naturalist, wrote about the gorilla: "It is unfortunate that the ape that, in some respects, stands nearest to man, never can be seen in adult state in zoölogical gardens; but we may as well accept the fact — because we cannot do otherwise." Since then a fair number of gorillas have reached adulthood in captivity, but since zoos usually lacked adult pairs, or, if they had them, kept the animals in separate cages, no breeding occurred. The Columbus zoo in Ohio was the first to breed a pair of gorillas in captivity. Sometime between April 6 and 8, 1956, the young female Christina copulated with the male Baron and conceived. After seven months of pregnancy, her ankles were swollen and she became very irritable. On December 22, in her ninth month, at 8:00 in the morning, she stood quietly in her cage on all fours. When she was checked by the keeper at 8:30, she had not moved, but at 8:50 the keeper found an infant on the floor still encased in its amniotic sac. Christina had made no sound, and her face had a dazed expression. The valuable infant was removed from the cage and successfully raised.

The zoo in Basel, Switzerland, has had the most spectacular success in breeding gorillas. The female Achilles was known to be pregnant in late 1957, but on March 29 she had a miscarriage and aborted a foetus about four inches long. In December of that year she was suspected of having conceived again. Then her breasts enlarged, her stomach protruded, and her weight increased by fifteen pounds. On June 7, 1959, she squeezed milk from her breast. The birth occurred on September 22 in the quiet of the night, 289 days after conception. When the keeper came into the cage in the morning, he found Achilles with her infant clasped to the chest. She cared for it tenderly, but she was an inexperienced mother. The youngster grew hungry and mouthed in search for the nipple, yet she made no attempt to aid it. Soon it whimpered, obviously hungry. The youngster was removed from the cage to save its life and raised in the home of Dr. Ernst Lang, the director, who published a charming account of his charge in *Goma, the Gorilla Baby*. He and his associates also studied the infant, giving us the first detailed information on the development of the gorilla from the time of birth. All too often zoos, especially in America, are no more than menageries; no attempt is made to keep even the most rudimentary scientific

Christmas at Kabara.

A blackbacked male peers at me from the cover of vegetation.

stand in the **Hypericum** *woodland. The summit of Mt. Gahinga is* *in the distance.*

Two females, a blackbacked male, and two large infants of group VI crowd onto a tree branch, thus obtaining a clear view of me.

notes, with the result that much valuable information is lost. To this the Basel zoo is a happy exception.

One month after the birth of Goma, Achilles was reunited with the male, and in early August, 1960, she conceived once more. On April 17, 1961, the keeper checked on her at 7:00 in the morning without perceiving anything unusual. When he returned half an hour later, Achilles had the newborn gorilla in her arms. The gestation period of this infant was 252 days, as compared to an average of 267 days in man. Achilles inspected her offspring carefully and licked its hands and feet, obviously less nervous than after the birth of her first child. The young was drinking from its mother's breast within a day, and, as the months passed, it grew and continued to thrive. Jambo was the first infant to be raised with its mother in captivity, showing that a zoo which treats its animals well, both physically and emotionally, can succeed in breeding and raising even the most delicate of creatures.

The most recent gorilla birth occurred in Washington, D.C. On September 9, 1961, after about 266 days of pregnancy, the young gorilla was born about 6:00 in the morning. As in all the other births, no one was present to observe the delivery. The female placed the newborn infant on the floor, showing little further interest. Nor was she concerned when the infant was removed by the keeper.

But worse was the treatment accorded an infant by its mother at the I.R.S.A.C. station near Bukavu, Congo, where a pair of mountain gorillas lived in a large enclosure. On October 26, 1959, the female interrupted her noon meal, lay down, and raised her right thigh. The head of the infant appeared. The amniotic sac ruptured within five minutes after she assumed this posture, according to the natives who were watching her. Then, half-sitting, half-lying, she took the head of the infant into her hands and pulled it out. She broke the umbilical cord, raised her offspring to her mouth, bit off one of its feet and a hand, and punctured its skull with a canine.

Zoo animals which give birth to a young for the first time in their lives frequently ignore the new addition to their cage or are afraid of it. To them a baby is new and strange, something with which they have had little or no previous experience. It is likely that gorillas and many other social mammals learn the appropriate way in which to handle their infants by watching other females with their infants. The animal psychologist Hediger relates an interesting example which shows how important tradition and imitation can be in the life of a primate. A female chimpanzee gave birth in captivity, but she did not know how to carry the infant on her back in the proper manner. Instead of having the youngster face forward, she had it looking in the direction of her rump. This was an awkward method, and, although she apparently sensed that something was wrong, she did nothing to correct the situation. Then she observed another female who carried her infant in the right position. Immediately she switched her young around to imitate what she had just seen.

We had set June 2 as our departure date from Kabara. We found it difficult to realize that our idyllic months in these mountains were coming to an end. Already we looked at the meadow and the forest and all the creatures with different eyes, more intensively and yet with sadness, consciously gathering the final impressions in a way that would make them forever indelible in our minds. Although we planned to return to Kabara, we knew that future stays would be short, at most a month or two, for the purpose of filling in gaps in my data, without time enough to become part of the mountains again.

On May 25, group VI appeared on the slope behind our cabin. When I saw Mrs. October, I was shocked to see that her eight-month-old youngster was seriously wounded. The skin on its rump was torn off and hung still attached between its legs, and the muscles were raw, with the naked bones sticking through. What had caused the mishap? Did a leopard take a swipe at the young or did an excited male gorilla bite it? The infant was very weak and it barely moved in its mother's arms, a sad contrast to its usual exuberant self. Mrs. October treated it solicitously, and she was greatly concerned about its condition. She did not permit it to ride on her back, but held it in her arms in such a way that no part of the wound touched her body. Once she looked at the injury intently and briefly picked at it with her fingers. Brownie, a female who had lost her own infant about two months earlier, came up and with pursed lips touched the face of the youngster as if kissing it. But when she attempted to repeat this ten minutes later, Mrs. October gently pushed her away. I saw the group for the last time on June 1. The infant was still alive, and its wound was somewhat crusted over. As we packed our belongings and waited for the porters to take us down into the lowlands, I wished with all my heart that the youngster would survive to play again on the slopes of its mountain home.

CHAPTER 9

"Am I Satyr or Man?"

When I began to study gorillas, I was tremendously impressed with their human appearance — they gave the superficial impression of slightly retarded persons with rather short legs, wrapped in fur coats. They stretch their arms to the side and yawn in the morning when they wake up, they sit on a branch with legs dangling down, and they rest on their back with arms under the head. In their emotional expressions too the gorillas resemble man: they frown when annoyed, bite their lips when uncertain, and youngsters have temper tantrums when thwarted. The social interactions between members of a gorilla group are close and affectionate, much like that of a human family, and their mating system is polygamous, a type for which man certainly has a predilection.

$\mathbf{O}$ne hundred years ago the explorer Du Chaillu first described a male gorilla "beating his chest in rage." Almost every hunter, traveler, and scientist who since that time has encountered gorillas in the wild mentions this striking display in which the animal rises on its hind legs and beats a rapid tattoo on its chest with its hands. But none of these observers noted that beating the chest is the climax of a complex series of actions, which more than anything else are typical of the gorilla and constitute the most exciting aspect of its behavior. The complete sequence, which is rarely given and then only by silverbacked males, consists of nine more or less distinct acts. At the beginning of the

216

display, the male often sits, tips up his head, and through pursed lips emits a series of soft, clear hoots that start slowly but grow faster and faster until they fuse into a slurred growling sound at the climax of the display. The hoots seem to generate excitement in the male, helping him to build up to the desired climax, much as natives use a drum in their frenzied dances. The gorillas have a look of great concentration on their faces when they hoot, and if another member of the group interrupts the even rhythm of the vocalization in some way, the male may stop abruptly and look around as if annoyed before continuing his display. Sometimes the male stops vocalizing briefly and plucks a single leaf from a nearby plant and places it between the lips, an act of such daintiness and seeming irrelevance that it never ceased to amaze me. The females and youngsters in the group know that the hoots and the placing of a leaf between the lips are preliminaries to rather vigorous, even violent, actions on the part of the male, and they generally retreat to a safe distance.

Just before the climax, the male rises on his short, bowed legs and with the same motion rips off some vegetation with his hand and throws it into the air. The climax consists of the chest beat, which is the part of the display most frequently seen and heard. The open, slightly cupped hands are slapped alternately some two to twenty times against the lower part of the chest at the rate of about ten beats a second. Gorillas do not pound their chests with the fists, as is often stated, except on very rare occasions. Chest beating is not at all stereotyped in its application, and the animal may slap its belly, the outside of its thigh, a branch, a tree trunk, or the back of another gorilla. One juvenile patted the top of its head about thirty times, and once a blackbacked male lay on his back with legs stretched skyward, beating the soles of his feet. Two females did not slap their chests directly, but rotated their arms, making their breasts flap in passing. While beating its chest, the gorilla often kicks a leg into the air.

Immediately after and sometimes during chest beating, the animal tends to run bipedally sideways for a few steps before dropping to quadrupedal position and dashing along. The male often slaps, breaks, and tears at anything in his path during the run, and this is dangerous, not only because of the violence of the act, but also because the male is not at all selective in what he swats. Any member of the group may be hit. One juvenile was picked up by a male and bowled down the slope. Even a man in the path of a running male may be swatted, as observed by Fred Merfield: "N'Denge was holding his gun loosely pointing downward, and was looking toward me, when a big male gorilla suddenly crashed out of the bush and swept him aside with a terrible full blow in the face." The grand finale of the display consists of a vigorous thump of the ground with the palm of the hand. The performer then settles back quietly, the display completed. It is a magnificent act, unrivaled among mammals, which Dr. Emlen likened to a symphony when he wrote (in the Transactions of the Philadelphia Academy of Science): "The hooting comes first, an effective introduction, rich in restraint and suspense to the human connoisseur of sonata and symphony. This leads with dramatic crescendo into the powerful climax of the display as the animal leaps to its feet, hurls leaves and branches into the air, and pounds out a resounding percussion on its massive chest. Then follows the crashing finale, a free-swinging run ending in a tremendous thump."

Infants display various acts at an early age. When only about four months old they rise shakily and very briefly on their hind legs and beat their chests, and at a year and a half they throw vegetation and place a leaf between the lips. Although all parts of the sequence, except for the hooting and possibly the kick, are given by the females as well as the males, the former display less frequently and less intensely. To what extent the display is inherited or learned, and how age and sex of the animal affects the behavior cannot be determined in the wild. Goma, the infant born at the Basel zoo, beat its chest and the ground without having had the opportunity to learn the act from other gorillas. This, and similar data, suggests that gorillas have an inborn tendency to beat something when excited.

A display like that of the gorilla poses some challenging problems of interpretation. The remarkable series of sounds, movements, and postures must be of survival value to the species or it would not have evolved or persisted. A clue to the function of the display and to its underlying cause can be obtained by noting the situations that elicit it. The most intense, prolonged, and diverse displays are given in response to the presence of man. Other situations include the proximity of another gorilla group or a lone male, an undetermined disturbance, displays by another member of the group, and play. The hooting, the bipedal stance, the thrown vegetation, the running — all are actions which draw attention to the animal and make it conspicuous. It seems to advertise its presence, to show off. Studies of many animals, especially birds, have shown that certain prominent gestures are of significance in signaling information to others of the same and even different species — they are a means of communication. Judging from the effect that the display of the gorilla has on other gorillas and on man, it appears to function in two ways: it is a communicatory signal, revealing, for example, that another group is in the vicinity, and it serves to intimidate other gorillas and human intruders.

But what is the motivation underlying this remarkable display? Intimidation and communication do not explain the cause. Why, for instance, does it occur in play and in situations where there is nothing to intimidate or communicate? The most general emotional term which encompasses all the diverse manifestations is excitement. During the display the gorillas find release for the tension which has accumulated in their system in an excitable situation. After the display, the level of excitement temporarily drops, and the animals behave calmly until a new accumulation of tension erupts in display.

The evolutionary derivation of the display is of particular interest, and my explanation of it is based on the concepts developed in recent years by Niko Tinbergen, Konrad Lorenz, and

other behaviorists. When two conflicting impulses, such as attack and flight, operate at the same time, the result is often a displacement activity, an activity which seems irrelevant and out of context in the situation at hand. A conflict of this type may produce preening in birds; but in gorillas it results in feeding, scratching, yawning, and the tendency to beat something. Natural selection may act on such displacement activities by enhancing their effectiveness as communicatory signals; they may become stereotyped and be incorporated into a definite display — they become ritualized. The fact that the gorilla often places a leaf between its lips suggests that this curious gesture may be a ritualized act of displacement feeding, and beating the chest may be the fairly stereotyped outcome of the tendency to slap something.

The gorilla shares various aspects of the display with other apes and with man. In Borneo I watched gray gibbons swinging from branch to branch by their arms, almost flying it seemed through the canopy of the rain forest. When they spotted me and grew excited, they hooted several times, faster and faster, until at the climax their sounds were of very high pitch and they ran bipedally along a branch. There is obvious similarity to the hooting, rising, and running of gorillas.

The gorilla's habit of throwing branches and other vegetation when excited is also found in the orangutan. I remember one evening toward dusk in Sarawak as my Dayak guide and I came across a female orangutan with a large infant both feeding on small green fruit in the forest canopy. The infant was about five feet from her when she saw us. Immediately she reached over, snatched the infant to her chest, and climbed from tree to tree for some five hundred feet, the youngster either clinging to her back or to her side. She grew very excited as we followed her. Three times she held the knuckles of her hand to her mouth and kissed them loudly, a sound which was followed by a *gluck-gluck*, resembling the gulping of liquid, and ended with a loud, two-toned burp. Then she peered down at us from a height of one hundred feet, a shaggy almost grotesque creature, black against the evening sky. Over a period of fifteen minutes she ripped off twigs and branches and hurled them at us. Several times she swung a branch like a large pendulum and at the peak of the arc closest to me she released it. The behavior of this female orangutan certainly seemed purposeful; at any rate, with the branches crashing down around me I had to jump nimbly at times to escape being hit. Dr. C. R. Carpenter has also observed that gibbons in Thailand, and howler and spider monkeys in Central America, break off branches and drop them in the direction of the observer. None of the gorillas I watched ever used branches and leaves as missiles, although vegetation sometimes inadvertently flew in my direction.

The chimpanzee exhibits nearly all aspects of the display sequence, although its behavior is, on the whole, not as stereotyped as that of the gorilla. Captive chimpanzees hoot, throw objects, slap floors, walls, and themselves, jump around, shake the bars of the cage, and stamp their feet when excited. Once when I surprised a male chimpanzee in the Maramagambo Forest of Uganda he raced along a branch and hid in a nest with only the top of his head poking inquisitively over the rim. We looked silently at each other until he suddenly slapped the edge of the nest and with an agility unheard of in a gorilla descended the tree. My most hair-raising experience in Africa was an encounter with a group of displaying chimpanzees in the Budongo Forest of Uganda. With Richard Clark, an anthropology student from Cambridge, I visited this forest for several days in early July, 1960, to observe chimpanzees. At dawn we crawled through the wet undergrowth in the direction of some hooting, barking, and gibbering chimpanzees that sounded like a conclave of maniacs. It had been light for half an hour when we reached the apes. Most of them were still in bed, squatting in their nests of branches anywhere from fifteen to ninety feet above ground. One juvenile left its nest and fed nearby on the olive-sized fruits of the *Maesopsis* tree. A female walked leisurely along a branch, but when she saw us she raced away through the tree and jumped twenty feet down into the leafy crown of a sapling, and from there to the ground. The others left their nests, hooting as they fled, and soon they were spread out in the distant trees, and on the ground where we could not see them. We followed the retreating animals, of which there were about thirty. Suddenly, as if by signal, all hooting ceased. The chimpanzees disappeared from sight, and we waited in the silent forest, scanning the tree tops and listening. Minutes passed. Without warning the hooting began again, this time all around us in the obscurity of the undergrowth, drawing closer and closer until the sounds seemed to come out of the earth itself. Not a single animal revealed itself, and this, coupled with the high-pitched screeches that appeared to erupt from the throats of a thousand furious demons, brought fear to our hearts. It was fear of the unknown, of being unable to do anything except wait. When the hoots reached their screaming climax, strange and new sounds reverberated through the forest — rolling, hollow, *bum-bum-bum*. Later we were to discover that the chimpanzees pound the hollow buttresses of ironwood trees much like an African beats a drum. The pandemonium subsided, and the chimpanzees retreated, leaving us thoroughly intimidated by their fascinating display.

Man behaves remarkably like a gorilla in conflicting situations. A marital squabble, for example, in which neither person cares to attack or retreat, may end with shouting, thrown objects, slamming doors, furniture pounded and kicked — all means of reducing tension. Sporting events, where man is excited and emotionally off guard, provide the ideal location for people-watching. A spectator at a sporting event perceives behavior that excites him. Yet he cannot participate directly in the action, nor does he want to cease observing it. The tension thus produced finds release in chanting, clapping hands, stamping feet, jumping up and down, and throwing objects into the air. This behavior may be guided into a pattern by the efforts of cheer leaders who, by repeating similar sounds over and over again with increasing frequency, channel the display into a violent and synchronized climax. Two of the functions of this display are communication with and intimidation of the opponent. Wherein lies the difference between gorilla and man?

What indeed are the differences between gorilla and man?

Am I satyr or man?
Pray tell me who can
And settle my place in the scale;
A man in ape's shape,
An anthropoid ape,
Or a monkey deprived of a tail?

When I began to study gorillas, I was at first tremendously impressed by their human appearance — they gave the superficial impression of slightly retarded persons with rather short legs, wrapped in fur coats. The gestures and body positions of gorillas, and for that matter also those of other apes, resemble those of man rather than the monkeys. They stretch their arms to the side and yawn in the morning when they wake up, they sit on a branch with legs dangling down, and they rest on their back with their arms under the head. The great structural similarity between man and apes has been noted repeatedly since the time of Linnaeus and Darwin, and it is for this reason that all have been placed taxonomically into the super-family *Hominoidea.* In their emotional expressions too the gorillas resemble man: they frown when annoyed, bite their lips when uncertain, and youngsters have temper tantrums when thwarted. The social interactions between members of a gorilla group are close and affectionate, much like that of a human family, and their mating system is polygamous, a type for which man certainly has a predilection. These and many other basic similarities are to be expected, for man and the apes evolved from a common ancestral stock of monkey-like apes which diverged, one line leading to the apes, the other to man. It must be assumed from the evidence of evolution that man became man by the slow accumulation of certain characteristics, he became man by degrees, but still retained in his mind and frame the stamp of his origin.

Because of the many similarities between the apes and man, scientists and philosophers have over the years been bedeviled by the problem of pointing to basic distinguishing characters other than minor anatomical ones. Some, like the famous French naturalist Buffon in 1791, glorified the mental capabilities of man by belittling those of the ape:

Thus the ape, which philosophers, as well as the vulgar, have regarded as being difficult to define, and whose nature was at least equivocal, and intermediate between that of man and the animals, is, in fact, nothing but a real brute, endowed with the external mark of humanity, but deprived of thought and of every faculty which properly constitutes the human species . . .

Others, like Lord Monboddo in his *Of the Origin and Progress of Language*, published in 1774, looked upon the apes almost as our equals:

The substance of all these different relations is, that the Orang Outang is an animal of the human form, inside as well as outside: that he has the human intelligence, as much as can be expected in an animal living without civility or arts: that he has a disposition of mind, mild, docile, and humane: that he has the sentiments and affections peculiar to our species, such as the sense of modesty, of honor, and of justice; and likewise an attachment of love and friendship to one individual, so strong in some instances, that the one friend will not survive the other: that they live in society, and have some arts of life; for they build huts, and use an artificial weapon for attack and defense, viz. a stick; which no animal, merely brute, is known to do. . . . They appear likewise to have some kind of civility among them, and to practice certain rites, such as that of burying the dead.

In more recent times man has often been defined by one criterion like tool-using and tool-making, the presence of language, or the possession of culture.

The use of tools is certainly not confined to man, as has often been pointed out, but is found in a variety of other animals, including the insects. One species of solitary wasp, *Ammophila urnaria*, holds a small pebble in its mandibles and uses it as a hammer to pound dirt into its nesting burrow. A Burmese elephant was observed to pick up a stick with its trunk and scratch its back. However, the mammal that uses a tool more frequently than any other except man is not an ape but the sea otter (*Enhydra lutris*). This large member of the weasel family was once almost hunted to extinction for its valuable fur, but in recent years it has again become abundant in the Aleutian Islands and along certain parts of the California coast. With Dr. K. R. L. Hall, of the University of Bristol, England, I studied the feeding and tool-using behavior of the sea otter at Point Lobos, California, in January, 1963. When feeding, the otters generally swam along the shore and around the reefs, diving and reappearing many times in a small area. After a dive, they popped to the surface, immediately rolled onto their backs and, holding a sea urchin, crab, or other item between the paws, began to eat. Occasionally an otter surfaced with a small black mussel and a fist-sized stone. The animal then rolled onto its back, placed the stone on its chest, held the mussel pressed between its stubby hands, and brought the arms down forcefully so that the hard shell of the mollusk struck the stone with a click. The otter banged the shell against this "anvil" many times in succession, pausing only briefly at intervals to see if the shell was cracked and the soft insides of the mussel exposed. One otter brought up 54 mussels in an hour and a half and banged the shells against a stone 2,237 times, truly an energetic tool-using performance. An otter sometimes uses the same stone to open several mussels in succession. Once, after an animal had fed on mussels, it dove with the stone, only to reappear with two crabs. After the crabs were eaten, it reached under one arm and placed on its chest the same distinctive stone it had used previously. Such retention of a tool suggests that sea otters may have the rudimentary ability to think about the relationship between objects even though one of these objects is not in sight, something for which apes have a very limited capacity. In general, sea otters seem to have the inborn tendency to handle and pound objects, and we suspect that the tool-using habit is learned by youngsters while watching their mothers feed.

The woodpecker finch, *Cactospiza pallida*, a drab little bird inhabiting the Galapagos Islands, provides a most remarkable example of tool-using. This finch resembles a woodpecker in that it climbs along tree trunks and branches in search of food, but, unlike the woodpecker, it has no long pointed bill with which to get at the insects. Instead, it picks up a cactus spine or twig, holds it lengthwise in its beak, pokes it into the cracks of the bark, and grabs the insects as they come out. Anthropologists point out that simple tool-using is of an entirely different order of mental activity than actual tool-making, and it is usually inferred that only man, and perhaps also the apes, have attained the level of cerebral development necessary to do this. The zoölogist R. Bowman made the following observation on the woodpecker finch: "One such bird was holding a spine about six inches long. Only about two inches of the spine protruded from the tip of the bill, the remainder passed along one side of the face and neck. Apparently the bird realized that the stick was excessively long, for it made an unsuccessful attempt to twist off approximately three inches of the spine by holding it with the feet." This was clearly a rudimentary attempt at tool-making.

The fact that captive monkeys and apes may pick up a stick and use it to rake in food has been noted many times. One chimpanzee, observed by the animal psychologist W. Köhler, fitted a small bamboo cane into a larger one to make a stick long enough to reach some bananas, a simple form of tool-making. Observations of this type on free-living primates are very rare. I never saw tool-using in wild gorillas, and the only reliable accounts are for chimpanzees. In Liberia, Beatty watched wild chimpanzees as they cracked open palm nuts. "He then picked up a chunk of rock and pounded the nut which had been placed on a flat-surfaced rock." The hunter Merfield observed several chimpanzees around a hole which led to a nest of subterranean bees: "Each ape held a long twig, poked it down the hole, and withdrew it coated with honey. There was only one hole, and, though for the most part they took turns at using their twigs, quarrels were constantly breaking out, and those who had licked off most of their honey tried to snatch the newly coated twigs." Jane Goodall, during her study of chimpanzees in Tanganyika, saw them push twigs into termite nests and eat the insects which adhered to them.

But, as Pascal once noted, "It is dangerous to let man see too clearly how closely he resembles the beasts unless, at the same time, we show him how great he is." Even though other animals share with man tool-using and to a minor extent tool-making ability, there still appears to be a wide mental gap between preparing a simple twig for immediate use and shaping a stone for a particular purpose a day or two hence. This was emphasized by the anthropologist Oakley when he wrote: "There is the possibility of gradation between these two extremes, perceptual thought in apes, conceptual thought in man; but it seems necessary to stress the contrast because one is apt to be so impressed by the occasional manufacture of tools by apes that there is danger of minimizing the gap in quality of mind needed for such efforts, compared with even the crudest tools of early man, which indicate forethought."

The most interesting archaeological discovery in recent years was the association of definite stone tools with the man-apes of Africa, the *Australopithecine*. In the Olduvai Gorge of Tanganyika, Dr. and Mrs. L. S. B. Leakey uncovered the skull of a man-ape with some simple chipped pebble tools, dating to the lower pleistocene, well over 600,000 years ago; and in South Africa similar tools were found in deposits that had also yielded the bones of man-apes. Two basic types of man-apes are known in Africa: *Australopithecus*, who was about four feet tall and weighed some fifty pounds, and *Paranthropus*, who was somewhat taller than the former type and weighed perhaps twice as much. Both walked fully upright through the savannahs which they inhabited and probably used their crude tools to kill and cut apart any small animal that crossed their path. The deposit at Olduvai Gorge, for example, contained the bones of frogs, rats, young pigs, and antelopes. But the remarkable thing about these tool-making man-apes is the fact that their heads are in many ways apelike and their brain capacity is only 450 to 750 cc., or no larger than that of gorillas. Brain size is, of course, not an accurate reflection of mental capacity, and judging from their systematic use of tools, the quality of the brain of the man-apes was considerably higher than that of gorillas.

Perhaps no aspect of ape behavior has more general interest than the way in which members of the group communicate with each other. Do apes have a rudimentary language, or do they merely emit a series of grunts and barks without meaning and without function? As I watched the gorillas over the weeks and months, a subtle change occurred in my thinking about the apes. At first I was highly impressed with their human ways, but there was something basic lacking, something that their brown eyes, no matter how expressive, could not convey, namely, a means of communication with each other about the past and the future and about things that were not immediately apparent. In other words, the gorillas lacked a language in the true sense of the word.

The apes seem to lack the tendency to vocalize for the sake of vocalizing, a trait which is so important in man. No infant gorilla ever babbled like a human baby. The gorillas had no interest in imitating sounds, in practicing with various combinations of sounds. The speech organs of gorillas and chimpanzees are perfectly adequate for talking. The failure of the apes to do so lies not in the anatomy but in the brain. Only with the greatest difficulty has a chimpanzee been taught to say the whispered approximations of "mama," "papa," and "cup." The apes are at the dawn of abstract and conceptual thought, but their neurological connections appear to be such that ideas fade away quickly. Thus, symbolic language, made possible through the ability to think in abstract terms, is the most unique feature of man. As Thomas H. Huxley wrote in an 1863 essay on *Man's Place in Nature*:

Our reverence for the nobility of manhood will not be lessened by the knowledge that Man is, in substance and structure, one with the brutes; for, he alone possesses the marvellous endowment of intelligible and rational speech, whereby, in the secular period of his existence, he has slowly accumulated and organized the experience which is almost wholly lost with the cessation of every individual life in other animals; so that now he stands raised upon it as on a mountain top, far above the level of his humble fellows, and transfigured from his grosser nature by reflecting, here and there, a ray from the infinite source of truth.

Or, as G. W. Corner noted more briefly: "After all, if he is an ape he is the only ape that is debating what kind of ape he is."

This is not to say that gorillas lack a way of communicating with each other, and that their method is not perfectly adequate for their simple mode of life. But the gorilla's ability to impart information to a neighbor is confined entirely to the situation at hand; there is no way to convey something that happened yesterday. On the whole, their signaling system is no more complex than that used by dogs and many other mammals. Gorillas coordinate their behavior within the group primarily by employing certain gestures and postures. For instance, a dominant male who walks away from a rest area without hesitation imparts not only the information that he is leaving but also the direction which he intends to take. In order to be groomed, a gorilla merely presents a certain part of its body to another animal. Each gorilla simply keeps its eyes on the rest of the group most of the time and does what the others are doing. Vocalizations, which feature prominently in our society, occupy a position of secondary importance in the gorillas; the animals are remarkably silent during their daily routine. I counted twenty-one more or less distinct vocalizations in free-living gorillas, with all but eight being infrequent. The apes grumble and grunt when contented, they emit a series of abrupt grunts when the group is scattered in the dense vegetation, they grunt harshly or bark when annoyed in some way, and they scream and roar when angry. These and other sounds appear to be mere expressions of the emotions: the sounds are not given for the purpose of communicating something; they are not symbolic. But the other members of the group have learned that certain vocalizations are given only in certain situations, with the result that many of the sounds have become definite signals. When, for example, the male suddenly roars, the other animals know that something potentially dangerous is in the vicinity, and they congregate around their leader. In general, vocalizations draw attention to the performer so that he can then impart further news through postures and gestures.

Although the number of basic vocalizations emitted by gorillas is fairly small, there is considerable variation in the pitch, intensity, and pattern of each sound. These variations greatly broaden the scope of the vocal repertoire, for the animals respond selectively to the sounds they hear. Their reaction depends not only on the sound but also on the condition under which it is given and the member of the group who gives it. For example, no member would mistake the deep, full grunt of a male for that of a female. One female had the tendency to scream loudly every time I arrived near the group. The others ignored her warning even when she was out of sight, indicating that they recognized her voice. Apparently she cried "Wolf!" too often. A sound may also have two meanings, depending on the situation. Harsh staccato grunts given by the leader when females quarrel causes them to subside. If, however, the male emits the same sound for no obvious reason, all members first look at him and then face the direction that occupies his attention.

One supposedly unique feature of man is his lack of precise instinctive responses to certain situations, a lack which has freed him from the strict biological control of most animals to let him to some extent chose his own destiny. The anthropologist Ashley Montagu expressed it as follows:

The development of intelligence increasingly freed man from the bondage of biologically predetermined response mechanisms, and the limiting effects they exercise upon behavior. In the evolution of man the rewards have gone not to those who could react instinctively, but to those who were able to make the best or most successful response to the conditions with which they were confronted. Those individuals who responded with intelligence were more likely to prosper and leave progeny than those who were not so able. If there is one thing of which we can be certain it is of the high adaptive value of intelligence as a factor in both the mental and physical evolution of man. In the course of human evolution the power of instinctual drives has gradually withered away, until man has virtually lost all his instincts. If there remain any residues of instincts in man, they are, possibly, the automatic reaction to a sudden loud noise, and, in the remaining instance, to a sudden withdrawal of support; for the rest man has no instincts.

Nobody, I think, would question the fact that much of man's behavior is the outcome of the culture into which he is born, that through learning, through social inheritance, he thinks and acts and makes those things which over the generations his particular culture has thought of as proper. But man's ability to learn from others and from experience, and to adapt his actions to conform with a predetermined set of rules, have merely masked the inherent aspects of his behavior. Man still possesses many instincts, and perhaps the most striking and unique one is the smile. An infant often begins to smile by the age of one month when the corner of its mouth is stimulated and when it hears various sounds. Later, during the second month, any smiling face, whether it is a person or merely a crude drawing of a face, will elicit a smile. A smile in the baby tends to evoke a smile in the mother, which in turn stimulates the baby to smile some more, a self-reinforcing mechanism which is very important in establishing the social bond between mother and child. Even infants born blind smile in response to certain stimuli. Although the smile as an important social pattern is unique to man, many other inherited types of behavior are shared with the apes — the tendency to throw and beat things when excited, to crouch down in submissiveness when threatened, and to develop a fear of strange objects at a certain age.

And yet the apes — and this is true of other animals — are not under the total grip of their instincts. Learning and tradition play an important role in their lives, a role which is difficult to assess with precision in the wild, because each youngster gradually and unobtrusively learns the things that help it to fit into its group and its environment. Knowledge of food plants, route of travel, the proper way to respond to vocalizations and gestures — these and many other aspects are undoubtedly part of the gorilla's tradition, handed down as a result of individual experience from generation to generation and constituting a rudimentary form of culture. The importance of tradition in animal society often becomes apparent only when a new trait appears. In recent years, for example, the blue tit, a European bird related to the American chickadee, acquired the remarkable habit of opening milk bottles on doorstops and taking the cream. This useful trait was apparently invented by a few tits in several localities, and it soon spread widely over western Europe.

The brain among the higher animals has evolved to increasing heights of complexity, growing more and more efficient and better capable of integrating the information received, and of becoming more aware of its surroundings as a whole. The ape brain seems to have evolved to or just over the threshold of insightful behavior, but it has not bridged the gaps that make it truly human. Why was *Australopithecus*, with the brain capacity of a large gorilla, a maker of stone tools, a being with a culture in the human sense, while the free-living gorilla in no way reveals the marvelous potential of its brain? I suspect that the gorilla's failure to develop further is related to the ease with which it can satisfy its needs in the forest. In its lush realm there is no selective advantage for improvement of manipulative skills like toolmaking, or of mental activity along the lines that characterized human evolution. There is no reason to make, carry, and use a tool if vegetable food is abundant everywhere and at all times and no preparation of this food is required beyond stripping and shredding it with the teeth and fingers. There is no selective pressure to try anything new or improve on the old. The need for tools and for new additions to the diet, like mice, antelopes, and other meat, is more likely in harsh and marginal habitats where a premium is placed on an alert mind and new modes of fulfilling bodily requirements. *Australopithecus* lived in such an environment, and man must have continued to evolve in a similar one. But the very existence of the gorilla, free from want and free from problems, is mentally an evolutionary dead-end.

CHAPTER 10

Uhuru

On the way back toward Rutshuru, several policemen threw a cordon across the road, but I merely bore down on them and they had to jump nimbly. We stopped briefly to tell Jacques the news from Rumangabo, then hurried on toward Uganda. At the edge of town, the road was barred by a log, and near it, outside a pub, milled several hundred Africans, many of them obviously drunk. As we stopped they ran toward us. To the first man that reached the car I gave a pack of cigarettes and asked him to move the barrier, which to our surprise he did. And before the rest of the crowd could engulf us we spurted away.

The history of Europeans in the Congo has been a brief one. In 1482 the Portugese explorer Diego Cao discovered the mouth of the Congo River, and in 1798 another Portugese penetrated into Katanga Province. Between 1870 and 1877 various explorers — Schweinfurth, Cameron, Livingstone, Stanley — entered the Congo, and in 1876 King Leopold II of Belgium formed the International Association of the Congo largely as a result of Stanley's urging. Belgium recognized the Congo Free State as the personal property of King Leopold in 1885. His reign caused an international scandal because of the mistreatment of the natives, and on November 15, 1908, Belgium assumed sovereignty over the colony.

For over thirty years the Congo remained a tranquil colony, seemingly isolated from the world around it. Belgian colonial policy was one of paternalism, concentrating on the material well-being of the masses with hospitals, housing schemes, and the regulation of wages. Such a policy, it was assumed, would insure a more contented population than would the granting of political rights. The evolution of the Congo was to have occurred in gradual stages, with mass education before the formation of an elite, and with local councils preparing the way for an ultimate democracy. Theoretically the plan was excellent, and the economic and social benefits which fourteen million Africans derived from it were readily apparent.

World War II and the years following it broke down the isolation of the Congo. African troops served abroad, and more and more villagers moved to the cities to escape compulsory labor in the mines, all of which tended to bring the people into contact with new ideas and concepts. Soon there was a growing realization that good government is not equivalent to self-government. Africans who had received a fair amount of education and who spoke French, the evolués, began to agitate against racial discrimination which existed in such matters as education, and they demanded "equal pay for equal work." Although Belgium made a definite effort at social integration, changes could not keep up with the rapid breakdown of paternalism and the growing political awakening. By 1957, there were demands for a definite program of emancipation, and several political parties were vociferous in their demands for freedom. In December, 1958, the Accra conference held that no African country should be a colony after 1961. These political developments excited African opinion and, together with overpopulation and appalling poverty in the large cities, created unrest and discontent. Paternalism perished on January 4, 1959, when the Africans in Leopoldville rioted after police tried to break up a banned political meeting. On January 13, less than one month before our arrival in the Congo, King Baudoin declared: "It is our firm intention without undesirable procrastination but also without undue haste, to lead the Congolese population forward towards independence and peace."

Life in the remote forests of the Kivu Province went on seemingly unaffected by the agitation for independence during 1959.

Over 80 per cent of the population was rural, far removed from the influences of the discontented in the cities. The villagers existed only in their tribal units, ignorant of the outside world and ignorant of their relation to it. If independence was conceived of at all, it was by the young men, and then only in individual and material terms of perhaps having cars to drive and new clothes to wear. The park officials, administrators, and planters to whom we casually talked thought of independence as something far in the future, almost as something abstract which need not be considered at present or prepared for.

In October, 1959, Belgium promised self-rule to the Congo in five years. Now the spirit of independence raced through the country like a brush fire. "Total independence, now, now, now," cried an ambitious politician named Patrice Lumumba in Stanleyville, and, when police tried to arrest him, his followers rioted for two days. At a conference with Congolese leaders in January, 1960, Belgium, seemingly in a fit of pique, suddenly declared that the Congo could have its independence by June 30 of the same year.

Kay and I learned little of these events until we descended from our mountain camp to Rumangabo in mid-January, 1960. Along the roads young men and children flashed the Churchillian V for victory sign with their fingers and shouted *uhuru*, meaning "freedom" in Swahili. Later, I inquired of Marc Micha if he anticipated trouble with independence. "But we have the Force Publique," he replied. "The soldiers will stop any trouble."

The Force Publique was created on October 30, 1885, to keep order in King Leopold's private domain. In 1960 it numbered about 24,000 Africans led by 1,100 Belgian officers. Many Belgians to whom we talked expressed their faith in this army, even though it had a long history of local rebellion against its white officers. On March 19, 1960, the Force Publique spoke out against Lumumba in a prophetic letter:

M. Lumumba was never in the Force Publique and will never be in it, and so how can he judge that there is no one in it capable of replacing the officers? . . . Dear Lumumba, beloved brother of the Whites, . . . do not forget that the Government is the Government thanks to the army. The hour of pushing us around like lambs is passed, and since you continue to repel us . . . we guarantee you the ruination of your powers and your country. . . .

The future of the Congo seemed inauspicious and we returned to Kabara on February 1, hoping to be able to complete the major portion of the study before Independence. In the ensuing months we heard that in Kasai Province the Lulua and Baluba tribes were fighting each other; that over 113 political parties were jockeying for power in the new government; that in Elizabethville miners were striking for pay raises and were dispersed by tear gas; that the government imposed a ban on the export of all funds, for in February alone over thirty-five million dollars left the Congo; that numerous Belgians were already fleeing; and that the settlers guarded the Goma airport day and night. It seemed wise to leave our isolated mountain home in early June and to work in nearby Uganda until we saw how Independence on June 30 would affect the Congo. The July 3 issue of the *Uganda Argus* reported:

The Congo's Independence Day passed in an unnatural calm in Goma and Kisenyi, with the streets deserted all morning and shops and restaurants closed.

Very few Africans were to be seen in Goma township, most of them preferred to stay in their own neighboring suburbs to celebrate.

Barbed wire barriers between Kisenyi and Goma were erected as part of the strict security measures, and a special pass was required for traveling between the two towns. From today onwards, a Customs post will be operating at the border between the Congo and Ruanda-Urundi.

Helicopters hovered above Kisenyi all day, while paratroops patrolled below in jeeps mounted with machine guns. Here too business houses were closed, and public holiday was declared until today. Street lights were kept on throughout the night.

No reports were received of refugees entering Uganda through Kisoro or Ishasha, where additional police forces have been stationed to deal with a possible influx. Bukavu was said to be calm.

The only incident reported was that a group of tribesmen from the Runi Forest emerged with bows and arrows at Butembo to celebrate "freedom" by hunting game in the Albert National Park.

Kay and I drove from Kampala to Queen Elizabeth National Park on July 9 to pick up Richard Clark, the Cambridge student with whom I had watched chimpanzees a few days earlier and who now wished to go on to Kisoro to see gorillas. Although on July 5 the first rumblings of a Force Publique revolt had come from Leopoldville, Kivu Province appeared calm. As we approached the border, long convoys of cars with Congo license plates passed us, heading into Uganda. Later, Frank Poppleton, the warden of Queen Elizabeth Park, confirmed our apprehension: the Force Publique near Rumangabo, Goma, and Rutshuru had rioted because the soldiers were not given expected promotions and control of the army. All whites were fleeing, and rumor had it that the Albert Park guards were killing the game on the plains. Richard and I left for the Congo the following morning, leaving Kay in the friendly care of the Poppletons. I hoped to retrieve some of my field notes which I had left behind in Rumangabo and, more than anything else, wanted to obtain an accurate picture of conditions in Albert Park and be of help to the officials if needed. Both Richard and I were also simply curious to find out what really was going on. The flight of the Belgians had all the earmarks of a panic reaction, with no one really being certain why they had fled and what the trouble actually was.

A tent city of the King's African Rifles had sprung up at the Uganda border post of Ishasha. We were told that forty cars, all from the small town of Rutshuru, had passed through yesterday, but that there was no further news. The road toward Rutshuru was deserted. No Africans walked along the road as they usually did, and the coffee plantations lay silent in the sun. At the edge of Rutshuru a jeep with rebel soldiers suddenly bore down on us and forced us to the curb. Four Africans, dressed in battle fatigues and steel helmets, jumped out, rifles in hand. I showed my American passport, and they searched the car for firearms. They were friendly, and when Richard asked for per-

mission to photograph them, they lined up and stood self-consciously erect. One pulled me into the picture, and we all stood there grinning broadly, surrounded by over two hundred Africans who had appeared as out of the earth.

Dr. Jacques Verschuren lived in a pleasant bungalow in Rutshuru, where we visited him. With his usual bubbly enthusiasm he greeted us, and then, with waving arms and many cries of "terrifique" and "mon Dieu," he told us how hectic it had been the day before when the Force Publique went from house to house searching for firearms. Several nuns and a gas station operator were the only other Europeans left in town. Of Rumangabo, only fifteen miles away, there was no news, but Goma was supposedly blockaded by rebels. Richard and I continued on to Rumangabo without mishap. There everything was quiet. The guards saluted as always. Baert, former assistant director at Rwindi camp, was obviously glad to see us. He was the only European left at Rumangabo, and he had had no outside news for several days. I picked up my notes and loaded some of my belongings into the car with a feeling of urgency.

On the way back toward Rutshuru, several policemen threw a cordon across the road, but I merely bore down on them and they had to jump nimbly. We stopped briefly to tell Jacques the news from Rumangabo, then hurried on toward Uganda. At the edge of town, the road was barred by a log, and near it, outside a pub, milled several hundred Africans, many of them obviously drunk. As we stopped they ran toward us. To the first man that reached the car I gave a pack of cigarettes and asked him to move the barrier, which to our surprise he did. And before the rest of the crowd could engulf us we spurted away. Near the border we had a flat tire, and, just after crossing, another. After that we limped along on the rim.

I later found that my decision not to tarry at Rumangabo had been a lucky one. Within half an hour after we left, rebel soldiers from the nearby military camp and several park guards beat up Baert and threw him into jail. They stood him against the wall and aimed their rifles as if to execute him, and then released him. The same evening a group of drunken soldiers visited Jacques and pointed a revolver at his face, and he had to talk to them for two hours before they would leave.

My most immediate responsibility was to find a home for Kay, who obviously could not return to the Congo. Through the generosity of Makerere College she was permitted to live in the girls' dormitory on the campus while I finished my work.

By mid-July over three thousand Belgian refugees had passed through Kampala on their way to Europe, and more continued to arrive. On July 19, after Kay was settled, I drove to Kisoro to obtain some recent news about the park. Various refugees, who had just left the Congo, advised me to wait two or three days, and I decided to climb to the summit of Mt. Sabinio. Reuben, two porters, and I wound our way through the fields, our feet kicking up the dry earth. The rough-toothed mountain was shimmering in the distance, and the cares of the world were far behind us. The Uganda Mountain Club had a hut by a meadow at the base of the peak, and we stayed there for the night. Von Beringe had passed near here in 1902 on his way to the discovery of the mountain gorilla on a ridge above us; and Lieutenant Weiss, who escorted the Duke of Mecklenburg Expedition, camped somewhere nearby in September, 1907, before being the first European to reach the top of Mt. Sabinio. The climb up through the heather and among the senecios was steep but easy, and we reached the summit in under three hours. Somewhere on these slopes Von Beringe had left a bottle with a note at the base of a bluff. To my knowledge no one has yet found it.

On my return to Kisoro, Walter Baumgartel told me that Count Cornet and Jacques had just passed through on their way to Kisenyi. Cornet had been urged to flee Rwindi in a radio call from Goma, and Jacques had been threatened by a drunken mob in Rutshuru. With their departure the last of the officials were gone from the park, leaving this magnificent nature reserve in the hands of the guards and poachers.

Kisenyi, where I had driven in search of Jacques, had taken on the aura of a military post. Paratroopers were lodged in the hotels, jeeps with mounted machine guns guarded the main approaches into town, and barbed wire and mines obstructed the main road to Goma. Yet on the surface the normal tempo of life went on. With nothing to do, the refugee women and children sunbathed on the beaches, played tennis, and sipped drinks in the open-air cafés. Any new rumor was passed from table to table, from hotel to hotel, creating uncertainty and confusion one day and hope the next.

Jacques and I went on to Goma over a back road. The town was surprisingly busy, except that many stores were closed and those that remained open had very little stock remaining on the shelves. Blue Congo flags with a large gold star in the center and six small stars, each denoting a province, waved from the buildings. At dusk all life ceased as the Europeans scurried to the safety of Kisenyi for the night. I went to the new African administrator to obtain a road pass, which I learned was needed to drive from one town to another. An African clerk sat in the waiting room beneath a row of photographs, each depicting one of the Belgian kings. The clerk examined my identification card for about an hour. His studies were interrupted by numerous Africans who came in to chat. To each newcomer I was formally introduced, and with all I shook hands as they arrived and again as they left. The Belgians are the most confirmed handshakers I have ever met, and the Congoleses have taken up this habit with a vengeance. After another hour, and considerable argument over my legible signature on the permit — which I finally had to change to an unintelligible scrawl to conform with the custom of the Belgians and hence Africans — I obtained my road pass.

Jacques and I were greatly worried over the future of Albert Park. Jean Miruho, the capable head of the Kivu Province at that time, had given assurance that the parks would be retained and that he wished the Belgian personnel to remain. But all park officials were now in Kisenyi, crating up their belongings for shipment to Belgium, and seemingly without any intention to return to their posts, even if conditions should improve.

N. L. Ochora, the African assistant warden of Queen Elizabeth Park, visited Albert Park on July 26–27, and his report describes what he found.

Mutsora: The general look of this place is like a place deserted several months ago. . . . Head Ranger Bikwetu Ferdinand claims to be in charge; so does the clerk.

Ishango: Below the camp the whole width of the Semliki was covered with fishing baskets and about 100 natives were attending them.

Rwindi: Around the Lodge there were about 30 rangers and 20 other people mocking and insulting the Manager who was under arrest. He had been brought back (from Goma) to show things to the two Head rangers. . . . The Head ranger of Rwindi came to me and sat in a chair, and several other people also came and sat down with their feet up saying it was very comfortable to sit on chairs in European buildings, and this was their first time . . . The Manager asked that he was going out to the warden's compound and back in a second. He got into the car, started it quickly and swerved in the opposite direction, and drove away at a speed never before seen in a park.

At a ranger post: . . . But one ranger broke in and said that if they were not paid on the proper day, they would go and kill the animals, and sell them for money.

We stayed near Kisenyi for several days, awaiting the arrival of a marauding Force Publique unit, which, according to rumor, was coming down from the north to Goma. In the evenings, I pulled my car onto the beach in Kisenyi and slept there. Sometimes, after dark, I slipped naked into the lake and floated on my back in the calm water. I looked at the sky, at the Southern Cross and, farther up, near the Milky Way, the Centaur. The black shadows of fruit bats moved on silent wings overhead to feed in the fig trees that lined the shore.

Jacques returned to Rutshuru by way of Kisoro on July 25, and I was to follow him the next day as soon as I had my car serviced. Rutshuru was calm when I got there. About 80 per cent of the planters had returned to harvest the red coffee berries, and even a few women and children were about. Jacques had been busy. With his characteristic vigor and courage he had gone to Rwindi and calmed the guards. After being leaderless for a week, they were ready to riot. We went to Rumangabo together to assure the guards there that they had not been abandoned and that we at least intended to continue our work in the area. In Kisenyi, the same afternoon, Jacques tried to persuade the park officials to return, for, if they failed to do so, the park might be destroyed. Even though conditions seemed safe, they refused to enter the Congo, and we sadly left them.

On July 13, the Security Council authorized the Secretary General to send United Nations troops to the Congo. On July 27, a contingent of fifteen Irishmen arrived by air in Goma and were promptly but briefly arrested by the Force Publique. The following day one hundred and fifty U.N. troops arrived, and all were now patrolling the border and the streets of Goma. The military situation had become rather confusing — and hilarious in a sad way. On the border stood Belgian paratroopers. Facing them were the Force Publique, the United Nations, a few Belgian officers who had remained behind, and all rode together in the same vehicles. No one had control over anyone else, which, I suppose, is referred to as a balance of power. At Rumangabo military camp, the Force Publique elected a new commander every week; as soon as he attempted discipline he was, in a democratic way, fired. Every so often a few rebels erected a road block and amused themselves for an hour or two by halting all cars, a rather harmless but nevertheless annoying pastime.

When after a week of relative quiet the Belgian park officials had not returned, Africans were appointed to their posts. The new head of Albert Park was Anicet Mburanumve, a twenty-three-year-old agronomy student, who, although untrained, was eager to learn and willing to take advice from Jacques, who became his unofficial advisor. On August 3, Anicet Mburanumve, Jacques, and I drove to Rwindi. Not a tourist had been there in over a month, yet laborers still constructed chairs for the new *rondavels* and game guides stood waiting to be hired. Life seemed to continue at its own momentum, for the present at least. So far, very little poaching had occurred in the park, and we were told that the fishermen at Ishango had all been arrested and given six-month jail terms. The future looked uncertain but not hopeless, and we felt that as long as there was money to pay the guards the park would survive.

Mburanumve assured me that I could continue my work unhindered in the park. I decided to spend four days visiting the active volcanoes, Mt. Nyiragongo and Mt. Nyamuragira. Only twelve miles separate Mt. Mikeno, on which gorillas are plentiful, from Mt. Nyiragongo, where gorillas have never been reported. The forest in the saddle between the two peaks is continuous but at one point only a mile wide. Why, I wondered, were gorillas absent from the active volcanoes now when surely they had crossed them in the past to reach the dormant ones? Did the volcanic eruptions drive the gorillas away, as some persons have suggested? Why do the gorillas fail to move back and forth between Mt. Mikeno and Mt. Nyiragongo as the elephants apparently do? During my visit I hoped to find some clues to account for the absence of gorillas, and, aside from all scientific endeavor, I simply wanted to climb the two peaks, drawn by the fiery lava and steaming fumaroles which I had been told existed on the summits.

On August 5, I took Andrea, two porters, and a park guard, and we followed a well-beaten trail toward Mt. Nyamuragira. This section of the park is open to tourists, who for a special fee may climb the peaks if accompanied by a park guard; trails have been cleared, and rest huts are spaced at convenient intervals. The sun beat down, and the lava rock lay sharp and twisted underfoot. Shrubs and brambles covered all but the most recent flows, and farther on, beneath a canopy of rather spindly trees, a layer of grass cushioned our steps. Nowhere grew the lush and succulent vegetation which is favored by gorillas. After four hours of hiking, first over fairly level terrain, then up the gentle slopes of Mt. Nyamuragira, we reached a rest hut. The following morning, the guard and I continued upward, out of the *Hypericum* woodland and onto the barren summit slopes sparsely covered with tufts of grass and stunted heather. We reached the rim of the crater, and I sat down with my feet dangling over the edge, looking over the crater floor some five hundred feet below me, over the sheets of black lava and the billowing smoke from the fumaroles rising perpendicularly into the cool morning air. The opposite walls of this huge crater were over a mile and a quarter away. We then traced the edge of the rim toward the south, where the wall ceased entirely, giving easy access to the crater floor. Gingerly we crossed the lava, careful to skirt those parts which sounded hollow underneath. The rock was contorted as if molded by the hands of a mad giant. Fumes of sulfur tainted the air, and white smoke rose through the ash and the jagged cracks in the lava. Here and there were dark holes from which the steam escaped with a sigh from the bowels of the earth. Ghosts live below us, according to legend, and when at night they grow cold, they fan the fires. We were in a world that was silent and dead, as it was in the beginning, when the earth was still young and life had not yet made its appearance. But then, on a mound of black lava, I found some yellow flowers. These bright and delicate blooms, full of life and beauty amidst the desolation around them, received all my admiration and remain my fondest memory of the crater.

We retraced part of yesterday's route before turning toward Mt. Nyiragongo, and camped for the night in a rest hut at the foot of the mountain. Elephants had emptied the water barrels, but we obtained enough water to cook our rice by squeezing the soggy leaves that had collected in the barrels. The climb up Mt. Nyiragongo was steep. Our path led through fairly lush forest, and at one point we followed the rim of an ancient crater heavily overgrown with *Hagenia* trees. Gorillas could easily subsist on the slopes of this mountain, but to reach it they would have to cross the barren fields of lava and scrubby forests that surrounded the base.

At an altitude of about ten thousand feet, at Baruta, stand two metal *rondavels*. Just before reaching them we came upon six

One female of group VI *squats in the crotch of a tree, her two-year-old infant beside her. The Kicker, a blackbacked male, sits on the sloping trunk. Brownie, a female, beats her chest.*

The Kicker stands and beats his chest.

In an attempt to intimidate me, a blackbacked male beats his chest.

A gathering of members of group VI. *Mr. Dillon* (right) *is holding a leaf between his lips just before rising to beat his chest.*

elephants pawing the soil of a seepage spring. We watched them for nearly an hour as they drank the water we needed for our evening meal. At last I shouted, and they lumbered off in single file. One of the huts contained beds and blankets to be used by tourists, the other, designated for the Africans, was without furnishings. I told the Africans that they ought to use the furnished hut for the night since I planned to sleep alone on the summit anyway, but tradition was too strong and independence too recent and they refused to do so.

My trail led through a stand of tree heather and then wound through a maze of lava blocks strewn wildly about the slope. I quickened my pace, straining upward, not really knowing what lay ahead but impatient to find out. Then I stood on the rim, without thought, looking at the tremendous hole in the earth that went down and down and at the white steam that mushroomed upward, and all around me was a sound like the growling of a monstrous dog. My mind refused to sort out the individual impressions. I slipped off my pack and sat on a boulder, gazing into the depths.

At my feet the somber perpendicular walls dropped four hundred feet to a broad shelf, followed by another drop, and another and another. Some twelve hundred feet below me bubbled a lava lake, often obscured by the steam that burst from two large vents just above it. The black surface of the lake heaved and bucked like a captive creature in a pit — a pit two-thirds of a mile in diameter. Jagged red fissures like wounds opened in the black pool, and spurts of molten lava squirted into the air.

Having climbed Mt. Nyiragongo, I had now reached the summit of all eight Virunga Volcanoes, to my knowledge the second person to have done so. In 1946, Earl Denman, known chiefly for having attempted to climb Mt. Everest on his own, had been the first to reach the top of all the peaks.

At dusk I crawled into my sleeping bag, which I had placed on a small level spot near the crater rim, and ate my dinner of sardines and crackers. Then I dozed, weary after my day's climb. I awoke after dark. A soft red glow suffused the sky as I crawled on all fours to the crater rim. I lay there and stared at a sight so beautiful that my heart wanted to cry out and waves of shivers moved over my back. Far below, the lake of lava shone a bright orange-red, and as the glow rose and spread outward it turned a delicate purple. I was an intruder, spying under the cover of darkness into a purple vortex at the center of the earth lying before me naked, its emotion laid bare.

The night was cold. Shivering, I crawled back to bed and gazed at the vault of heaven. Soon an icy wind tore at my body and moaned among the boulders and hollows. A full moon rose and poured its silver light over the bleak slopes. It sailed nearer and nearer, the black depths behind it growing more and more distinct. Across the valley rose Mt. Mikeno, dark and still, to the stars. I was alone in space, adrift in the sky. Never before had the insignificance of my being struck me so overwhelmingly. On one side a fiery red glow suffused my body, attesting that the earth is not a stable abiding place; on the other, the moon caressed me with a soft light from the immensity of the cosmos.

For two months now the slopes of Mt. Mikeno and Mt. Karisimbi and all the others had been left unattended, and I feared that the feeding grounds of the gorillas had been usurped by the cattle, which, I knew, grazed illegally on the mountain slopes. On August 10, the porters and I left Kibumba for Kabara. The walk was the saddest I had ever taken. Cattle trails criss-crossed the bamboo, and, higher up in the *Hagenia* woodland, the lush green undergrowth was gone. Hundreds of hooves had churned the fragile soil, which looked like a plowed field. Cow dung lay splattered everywhere, and the musty odor of the beasts permeated the air. We passed three herds, and as I watched the dull creatures ravenously consuming the beautiful foliage, fury mounted in me. With a park guard I raced ahead of the porters. A cow stood on the trail and I brought my heavy walking stick down on her neck. Her knees buckled briefly before she lumbered off, lowing and lowing. Further on, at the base of a large tree, stood a tepee-like hut of Watutsi herders. With a bellow the guard and I charged forward, and two men and a boy fled naked into the forest. The guard burned the hut down and broke the spears which the Watutsi had left behind. Finally we reached Kabara. The interior of the hut was wrecked: the furniture smashed, the windows gouged out, the doors ripped off the hinges, the mats burned. The remains of duikers floated in the water barrels, polluting the water. I was glad that Kay was not there to see what had happened.

I bought two spears from the porters for protection before sending them home with a note to Mburanumve asking for guards and rifles to once and for all rid the park of the invading cattle. Andrea, the guard Bon Année, and I were cleaning up the hut when eleven Watutsi armed with long spears stepped from the forest onto the meadow. They stood there silently, close together, scowling. I told them that if the cattle were not gone by tomorrow the beasts would be shot. The Watutsi spoke in low tones among themselves, laughed at me, and abruptly disappeared. Their behavior seemed ominous. We loosened a board at the back of the cabin to provide an escape route if necessary. We slept together, the door barricaded, our primitive weapons beside us.

Both Andrea and Bon Année were Bahutu, enemies of the Watutsi, and when I wanted to scout around the forest the following morning I had to go alone, for I did not dare to leave either of them alone. I roamed upward to the Rukumi meadow and downward to Rweru, keeping off the main trails, without finding cattle. Perhaps my threats had helped, for during the night the Watutsi had moved their livestock from the vicinity of Kabara.

At the base of a tree, I stumbled on the remains of an adult male gorilla. He lay face down, arms askew, in a bower of vines that draped from the leaning trunk, as death had overtaken him.

The whole skeleton was beautifully preserved and eaten clean by maggots. When I examined the bones, I found that the left forearm was severely deformed, caused apparently by a fracture of the radius bone which subsequently healed unevenly.

In the ensuing two days I searched for and found my gorillas again — on the steep slopes and in the nettle fields where cattle had not ventured. I renewed my rapport with the animals: the Eskimo was still in group V, and the incomparable Junior in group IV. In general the animals were as friendly as ever, and I felt certain that they remembered me. The psychologist Yerkes found that a captive mountain gorilla still recognized him after an absence of one year. It was peaceful to sit with the gorillas again, propped against a tree trunk, the forest and mountains so still all around, to watch the apes behaving quietly and with dignity among themselves and with tolerance toward me. But the tranquility of the past I could not recapture, for my attention was divided, attuned both to the gorillas and to the possible stealthy approach of a Watutsi.

On the evening of August 13, five park guards and two policemen arrived at Kabara fully armed with rifles and with the authority to shoot and confiscate any cattle in the park. The Belgians had been fearful of employing severe measures to save the park from destruction, but, to my delight, Mburanumve had the courage and foresight to deal with the cattle problem in the only emphatic way. We found the cattle near the Rwanda border. Running rapidly through the undergrowth, the guards surrounded a large herd. Two Watutsi broke through the cordon, escaping without harm. Senkekwe, one of the guards, calmly lifted his rifle and shot a cow in the neck. She dropped and kicked, then lay still. The others too opened fire at the densely crowded herd. Some fell, others stood lowing as the bullets ripped through them. A guard dropped his rifle and plunged his long knife to the hilt again and again into the stolid cows, until his arm was red with blood and his hate against the Watutsi vented. I counted twelve dead cows, and there may well have been more. The guards hurriedly rounded up 56 head and with much shouting drove the reluctant herd almost at a run back to Kabara and on down the trail to Rumangabo.

The Watutsi watched our meadow from the forest, remaining hidden with only their straying dogs revealing their presence. We stayed in the hut, waiting. I realized that my work was finished at Kabara, and now my immediate concern was to get us safely off the mountain. Mburanumve came to our rescue on August 16 by sending porters up to fetch us. As we left, I did not look back at the hut and the trees and the mountains behind me, for I wanted to remember them as they had been, not as they were now, defiled by cattle.

With further work in the Virunga Volcanoes unsafe for the present, I decided to revisit the Congo basin, the Utu region, to fill several important gaps in my knowledge of the area and the gorillas in it. No one in Goma seemed to know for certain whether or not the road to Bukavu was safe or even open, but taking a chance, Andrea and I reached the I.R.S.A.C. station without trouble. The director of the station had fled, and most of the women and children were safely in Rwanda, but otherwise research continued under the leadership of a three-man committee, which included the mammalogist Urs Rahm.

Charles and Emy Cordier now lived nearby in the home of a planter, Michele de Moevius, who was also a dedicated ornithologist. During the uprisings the Force Publique had rounded up all whites around the towns of Kabunga and Walikale and held them prisoner, but the Cordiers escaped capture by hiding in the forest. While I prepared for my journey to the lowlands, De Moevius offered me the hospitality of his home. Earlier he had been threatened with death by the laborers on his farm, and some months later a mob came and seriously injured Emy Cordier, but during my stay everything was calm. I needed three hundred liters of gasoline for my trip, but gasoline was rationed and it took me several days of scrounging to obtain it.

My trip through the lowlands with Andrea and Mathias, one of Charles' African helpers, went backward through time. The mining camps were shut down, and in all the many miles between Kabunga and Kasese I found only three Europeans, keeping mostly to their homes. Work on the dirt roads had ceased, and already the jungle was reclaiming these narrow strips of civilization: grass had pierced through the hard-packed soil, and the torrential downpours had carved miniature canyons in the roadbeds. The foundations of the wooden bridges were rotting and snapping under the onslaught of the turbulent rivers. Independence had not really come to the villagers. Rather, their life had reverted to what it was in precolonial times before Europeans had established their tenuous foothold. Here there were no shouts of *uhuru*, and Lumumba and Kasavubu meant nothing to the older generation. When I squatted on the ground, back against the mud wall of the hut that was my temporary home, dirty, unshaven, face red-splotched from the bites of flies, with no companions except Andrea and Mathias, I felt remote and for the first time somewhat lost in the immensity of the forest.

The rains came daily, sheets of plunging water that left the forest sodden and steaming and the roads almost impassable. By September 4, after a little over a week in the lowlands, I knew that the weather had defeated me. With the bridges collapsing, it was wise to leave before escape by car became impossible.

Later, at Rutshuru, Jacques told me that Albert Park was still intact. Only the other day he had swum across a river to kill the dogs of a poacher who had caught two wart hogs. The Rumangabo guards had confiscated about fifty cattle in the volcanoes. The Rwanda side of the park, still under Belgian rule, continued to be without authority, abandoned long before independence by an indifferent administration which made no effort to prevent incursions of cattle, poachers, and woodcutters. For two months now Jacques had been the driving force behind the efforts to save Albert Park, living in uncertainty, risking his life to preserve the natural heritage of the Congolese and the past efforts of the Belgians. Yet in all that time he was completely ignored by the park headquarters in Brussels, receiving not a single word of encouragement or even a direct acknowledgment of his letters. No wonder Jacques was depressed and weary. We decided to escape to the freedom of the mountains, to visit the gorillas at Kabara for a week.

We stayed at Kabara from September 11 to September 17, a week I thoroughly enjoyed, even though the weather was as usual rainy and damp clouds had descended to the level of the meadow. Daily we roamed the slopes, up toward Mt. Mikeno, to Bishitsi, and to Rukumi, saying farewell to the places I loved. And daily we observed gorillas, photographed them, and made tape recordings of their voices. There were no cattle, "those awful animals" as Jacques called them, around Kabara now. A new growth of herbs covered much of the ground, and soon, if left undisturbed, the last scars of the cattle invasion would be healed. In the evenings, huddled around the stove, Jacques told me of his work in Albert Park over the past few years. Of special interest to me was his tale of the *masuku*. At about ten spots in the park, usually at the edge of a lava flow, deadly gases escape from the ground. The *masuku*, as the natives call them, consist of shallow depressions overgrown with grasses and sedges and surrounded with trees and shrubs, innocuous-appearing places. The gas emitted is carbon dioxide in such high concentrations (over 40 per cent) that any animal exposed to it becomes intoxicated and dies of anoxia, the lack of oxygen. Attracted by the lush grass in the *masuku*, elephants, hippopotami, baboons, buffalo, and forest pigs enter only to die. Hyenas and other scavengers arrive and succumb before they can make a meal of the cadavers. About fifty years ago, according to the local natives, a band of some ten Warega tribesmen passed a *masuku*. Attracted by a dead elephant, they started to feast, only to succumb to the gas, adding their bodies to the many other skeletons that littered this natural graveyard.

We spent our last days with the gorillas of group VII, the group I had known more intimately than any other. I remember my final visit with them as we stood on a slight rise, watching the animals spread over the opposite slope, feeding leisurely. I suspected that once I moved from their sight I moved from their minds; they would go on feeding and resting and sleeping as they had always done, living in the present, without past and without future. I realized that my months in the mountains with them would be one of the happiest memories of my life, that mental images of the majestic males and plump females would come to me again and again in the coming years at the most unexpected moments. And I was sad that I could not communicate my affection and admiration to these gentle beasts before I left. I wanted to thank them for all they had taught me about themselves, about the ways of the forest, and about myself. All I could do was wish them luck and a free life of roaming about the mountains undisturbed by rapacious man and his cattle. I left them as I had found them a year earlier, sitting quietly, peaceful, content, watching our retreating forms as we disappeared over the crest of the ridge.

I decided that further work on the gorillas in the Congo was at that time not worth the effort and risk. The lowlands were impassable, the Itombve mountains too isolated, the Virunga Volcanoes too dangerous. A wild rumor had it that the former Belgian commander of the Rumangabo Force Publique camp was setting up artillery at Kabara to destroy the surrounding towns, and bands of Watutsi still roamed the forest. The park was intact, and although the future remained uncertain, there was nothing I could do to help Jacques. When I returned, Mburanumve had his administrative duties well in hand, and his thoughts for the future were well expressed in a manifesto circulated on November 9, 1960.

Let us remember that the Park Administration must not be concerned with any political point. The danger against which we must keep constant vigil is that there be no mixing of any organization and its administration, even if independent, with politics and selfish ends. The Parks are national heritages belonging to all those who form a nation — to the whole population, not to a single clan or tribe pretending to have rights of first occupancy or of land-ownership.

Thus there would be a continued union of the Congo Parks without any separation and later the internationalization of these Parks with the friendly world. Thus would be better realized our three goals: integral conservation of nature, co-operation in scientific research, and expansion of tourism, by which we are known throughout the world.

If finally our own [flora and fauna] were to disappear, and by our own fault, what would be international opinion of the Republic of the Congo?

On September 24, I left Rumangabo. I passed through Kisoro and said goodbye to Walter. And then I followed the winding road up out of the valley, higher and higher, until I reached the top of the escarpment. There I stopped and briefly looked back at the mountains. After that I dipped over the crest and continued on.

EPILOGUE

Once again I climbed upward, tracing my old route along the edge of the Kanyamagufa Canyon toward Kabara. In September, 1960, when I descended this mountain path, I had done so with the feeling that I would never return, that a phase of my life was irrevocably over. But the towering mountains and silent forests had cast a spell over me and I longed to return. So when an American magazine asked for some good color photographs of gorillas, I willingly offered to take one of their photographers to Kabara. Now, on August 11, 1963, three years after my departure from the Virunga Volcanoes, Terry Spencer and I followed the line of porters through the bamboo and up into the *Hagenia* woodland.

At times I looked ahead with hesitation, even trepidation, afraid that the forest in which Kay and I had spent an idyllic year had somehow changed; perhaps it had been devastated by cattle which had invaded the national park. To my delight, I saw no cattle, nor did I find recent signs of their presence: the forest spread peacefully and undisturbed up the slopes. The Kabara meadow was little changed. The marker on Carl Akeley's grave was buckled and shattered, apparently by the ponderous foot of an elephant. The hut still stood, ugly as ever, built without regard for the beauty of the spot. The walls of one room and one shed were burned away, the result of a fire, but two of the rooms remained usable, and into these we moved. Our old iron stove and one table were the sole pieces of furniture remaining. Later, as I watched our houseboy Andrea wash the dishes, and the smoke poured out from all the seams of the stove,

254

and rain drummed on the roof, it seemed as if nothing had changed at all.

In the morning we went into the forest between Kabara and Bishitsi in search of fresh gorilla spoor. Within an hour we discovered some nests of the previous night. Which group was it, I wondered, as Spencer readied his cameras. Would I be able to recognize the animals after an absence of three years? Would they recognize me? The gorillas had already begun their midmorning rest as we approached carefully to within a hundred feet and erected the tripod and camera. With a curious blend of cautiousness, boldness, and seeming disinterest, the gorillas stayed near us for over four hours, but we rarely glimpsed the adults. They remained in the dense undergrowth, occasionally peering at us from the cover of a tree trunk. They moved off in the afternoon, at first feeding slowly, then racing for the steep slopes of Mt. Mikeno and climbing to an altitude of over eleven thousand feet. There we found them the following morning. This time they fled at our approach, and we knew that further pursuit would be fruitless. The group consisted of one silverbacked male, two blackbacked males, seven females, three juveniles, and at least three infants. From glimpses of the big male and one or two females, I felt confident that this was my former group II, but I could not be certain.

The next day, within a few minutes after leaving Kabara, I spotted a gorilla in a nest at the base of Mt. Mikeno. It was just after seven, and all the members of the group were still asleep. A female sensed my presence, for she sat up and began to look around nervously. After fixing me with a brief, hard stare, she dashed some hundred feet up the mountain. Awakened by her actions, the others followed her in a straggly line. When the silverbacked male stopped to look back at me, I knew him immediately. He was the leader of group VIII. I left the animals to feed in peace and returned with Spencer later in the morning. I scanned the faces of the three females sitting placidly on the sloping trunk of a tree about ninety feet from us. With great pleasure I recognized one animal as an old and particular friend. She was a round-faced female with a rather puckish expression and manly de-

TABLE 2. THE COMPOSITION OF THE SAME GORILLA GROUPS
IN 1960 AND 1963

GROUP	Date Observed	Silverbacked Male	Blackbacked Male	Female	Juvenile	Infant	TOTAL
VII	Sept. 16, 1960	1	2	7	4	7	21
	Aug. 19, 1963	1	1	8	3	4	17
VIII	May 24, 1960	1	2	8	3	7	21
	Aug. 14, 1963	1	0	5	2	2	10

meanor. She had had no infant in the period November, 1959, to May, 1960, and she lacked an infant now. The past three years had not been kind to her. She had become an old woman, slightly stooped and quite gray on head and shoulders. And yet her boldness and curiosity had not changed, and as she squatted there watching me intently, I wondered if in the recesses of her mind there was a glimmer of recognition. Another former acquaintance was Mrs. V., who in September, 1959, had given birth to an infant. This youngster had apparently grown up to take its place in the group as a juvenile, and Mrs. V had given birth again in July, 1963, about a month before our present visit. Some familiar faces were missing. Mrs. Twist, the female with the broken jaw was gone, as was a juvenile with a cut nose. Something drastic had happened to the group, for it numbered now only ten animals as compared to twenty-one when I last saw it (see Table 2). Did a blackbacked male grow up and leave with some of the females to form his own group? Or did some members switch their allegiance to another male when two groups met in the forest?

The following four days were frustrating. Clouds hung low over the mountains and heavy showers drenched the forest. Group VIII fled up the slopes of Mt. Mikeno, choosing the brushiest and steepest route. Whenever we attempted to photograph, either the fog moved in or the animals were in the gloomy depth of a canyon where the light was too poor for pictures. After two days we abandoned this group and hoped for better luck with another group which I had found farther down the mountain in the range formerly occupied by group IV. But this group, like all the others, traveled upward as soon as it perceived us, ascending without pause from the bamboo zone to the zone of giant senecios. After a long and discouraging climb through fog and rain, I fleetingly caught up with the animals, still hurrying along across deep ravines and slippery slopes. I gave up the chase reluctantly without having been able to learn anything definite about the group except that it was large, containing at least twenty animals. Even our raven friends failed to revisit Kabara. One afternoon two birds sailed by overhead, leaving me to wonder if it was the same pair that had once fed at our doorstep.

A park guard and I then searched for gorillas in the forest toward Bishitsi and within three hours had located the fresh trail of a fourth group. But by the time the guard had brought Spencer to the site and I had traced out the meandering route of the animals, the sun had disappeared and the afternoon was well advanced. Yet we returned home with confidence, feeling certain that the next day would bring success. The morning of August 19 was sparkling and clear, a morning of perfection rare at Kabara. Spencer finally obtained the gorilla pictures he desired: pudgy females with their youngsters, and members of the group clustered around the huge male. And I was delighted to once again watch group VII. The silverbacked male, the Climber, was still the boss, but his head, neck, and shoulders were now quite gray, and he somehow seemed less powerful and dynamic to me. Equally dramatic was the transformation of Shorthair, whom I once knew as a rather small blackbacked male. Three years had changed him from a teenager to a young adult. His body had filled out and grown tremendously in size, his voice had deepened, and his back was silver. But his irascible roars showed that at least in temperament he had not changed. Mrs. Gnath had aged too, although her stern and domineering expression remained. I was surprised to find her carrying a large infant about a year and a half old, for when I left in September, 1960, she already had an infant only a year old. Thus she either conceived when the first infant was less than two years old, which seems unlikely, or it died and she became pregnant again soon after. We photographed various members of the group on the following two days, but then desisted in our pursuit of them, for they sought to elude us on the cliffs of Mt. Mikeno.

All four of the gorilla groups fled for a mile or more immediately after they met us, and each retreated up the almost inaccessible slopes of Mt. Mikeno. Although still amiable, curious, and stoic, the animals seemed to make a definite effort to escape as far as possible from the area in which they had been disturbed. This reaction represented a drastic change in behavior, for no group responded in this fashion during my earlier study. I suspect that frequent and perhaps adverse contact with poachers had caused them to be wary.

The porters came to carry our equipment back down the mountain on August 22. We left Kabara satisfied with our brief stay. Spencer had taken a unique series of color photographs of gorillas. I had seen the apes on each of our ten full days at Kabara, and I had been able to identify two of my former groups as well as six old gorilla acquaintances. It was also of interest to see that each of the groups still occupied the same home range as three years ago. And, above all, I was happy and relieved to note that so far the park authorities had effectively preserved the Kabara area from the agriculturalists and pastoralists, and that the gorillas seemed to be as abundant as ever. Even the fact that the gorillas were now shyer than they had been promised well for their continued existence, for only by adapting to changing circumstances can they hope to survive man's persistent onslaught on their mountain home.

SELECTED READING

The following titles will serve as additional reading on topics which are only mentioned briefly in this volume. My scientific report on the gorillas gives detailed documentation to all general statements made about the apes in the foregoing pages.

AKELEY, M. J. 1929. *Carl Akeley's Africa.* New York.
> A pleasant account of the Virunga Volcanoes by an expedition which made the first scientific attempt to study the mountain gorilla.

BAUMGARTEL, W. 1960. *König in Gorillaland.* Stuttgart.
> A lively account of life at the foot of the Virunga Volcanoes by the man whose hotel is unofficial "gorilla headquarters" in Central Africa.

BOURLIÈRE, F., and J. VERSCHUREN. 1960. *L'écologie des ongulés du Parc National Albert.* Fasc. 1. Inst. des Parcs Nat. du Congo Belge, Brussels.
> An excellent summary of the various habitats in Albert Park, and the only detailed account of the large game animals of this region.

CORNET, R. 1955. *Maniema.* Brussels.
> Fine descriptions of the various expeditions that penetrated to the headwaters of the Congo River and explored the vast Maniema Forest.

GREENE, J. C. 1959. *The Death of Adam.* Iowa State University Press.
> A good discussion of the emergence and development of evolutionary thought.

HARRISSON, B. 1962. *Orang-utan.* London.
> A pleasant book which summarizes our scant knowledge of this rare ape and describes the enjoyment and problems of raising infant orangs in the home.

Huxley, J. 1960. *The Conservation of Wildlife and Natural Habitats in Central and East Africa*. UNESCO.

An excellent and concise summary of the problems besetting the game animals in Africa.

Lang, E. 1963. *Goma, the Gorilla Baby*. New York.

This charming book relates the early life of Goma, the second gorilla infant to be born in captivity; excellent photographs.

Merfield, F., with H. Miller. 1956. *Gorilla Hunter*. New York.

A good general account of the lowland gorilla by a hunter.

Montagu, M. (ed.). 1962. *Culture and the Evolution of Man*. Oxford University Press.

A useful compilation of articles on various aspects of man's evolution.

Moorehead, A. 1959. *No Room in the Ark*. London.

Pleasant ramblings through Central Africa, including a visit to the Virunga Volcanoes.

Moorehead, A. 1960. *The White Nile*. New York.

An excellent account of the various explorers who penetrated Central Africa.

Schaller, G. 1963. *The Mountain Gorilla: Ecology and Behavior*. University of Chicago Press.

The behavior of the gorilla is documented in detail.

Slade, R. 1960. *The Belgian Congo*. Oxford University Press.

A concise account of some of the factors which contributed to the unrest prior to and following Congo independence.

Yerkes, R., and A. Yerkes. 1929. *The Great Apes*. Yale University Press.

A fine compendium summarizing all knowledge about ape behavior in the wild and in captivity to the year 1928.